ENGLISH RUGBY 101

A POCKET GUIDE

IN 101 MOMENTS, STATS, CHARACTERS AND GAMES

JOHN GRIFFITHS

POLARIS
PUBLISHING

This edition first published in 2019 by

POLARIS PUBLISHING LTD
c/o Aberdein Considine
2nd Floor, Elder House
Multrees Walk
Edinburgh, EH1 3DX

Distributed by
ARENA SPORT
An imprint of Birlinn Limited

www.polarispublishing.com
www.arenasportbooks.co.uk

ISBN: 9781909715806
eBook ISBN: 9781788851817

British Library Cataloguing-in-Publication Data
A catalogue record for this book is available on request from the British Library.

Designed and typeset by Polaris Publishing, Edinburgh

Printed in Great Britain by MBM Print SCS Limited, East Kilbride

Photos courtesy of:
Inphophotography
Getty Images
John Griffiths' archive
Arena Sport archive

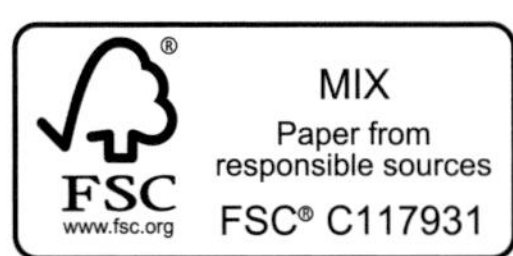

Acknowledgements

My thanks go to the 1405 rugby players capped by England since 1871, as well as to the legion of legendary reporters and writers who have recorded their deeds. History, though, is an iterative process; details of the past are constantly challenged in order to converge on an accurate record of events. I am therefore also indebted to the veteran English rugby researcher Tim Auty, to Phil Atkinson and Philip Tarleton of the Rugby Memorabilia Society, and to Richard Steele and Phil McGowan at Twickenham's World Rugby Museum for sharing primary evidence that challenges some facts relating to early international matches. Several of their rugby nuggets are embedded in this compendium. Finally, I should like to thank the rugby-mad Peter Burns of Polaris Publishing whose enthusiasm for this – and many other oval projects – has been infectious. There can never be too many books about rugby union.

Introduction

What is it about English rugby that unites other nations' fans, albeit in good humour, against it? This pocket guide provides some answers as to why neutrals invariably support anyone facing the men in white.

England devised the handling game and framed its first laws, and England Rugby, the rebrand of the original Rugby Football Union, is the oldest union in the world. Which rule-taker wouldn't relish the opportunity of teaching their rule-maker a lesson, or putting down the senior union?

But this distillation of 101 facts and stats, famous characters and matches sets out the many achievements – as well as some of the disappointments – of the national side, from taking part in the sport's oldest international fixture (against Scotland in March 1871) to the highest score-draw in Test history (against the same adversary) that brought down the curtain on the 2019 Six Nations.

It is a collection that shows England has a rugby history to be proud of – more Championship titles, Triple Crowns and Grand Slams than any other nation, and three Rugby World Cup final appearances since the tournament began in 1987. It is a record to rightly make others jealous.

The statistics go up to 31st March, 2019

FAMOUS FIRSTS

Rugby's first national union

The Chinese, Greeks and Romans all had words for a pastime that passed as a rudimentary form of football, and records show that mass brawls involving rival parishes mauling for possession of a ball took place on Shrove Tuesday holidays in Britain until the early 1800s. But it was the rise of the public schools during the Victorian era that accelerated the laying down of rules that transformed mob football into more refined versions. Different schools adopted different rules but dating the exact origins of the game that evolved at Rugby School and was characterised by handling and scoring goals by kicking the ball high over a crossbar is blurred by conspiracy theory and myth. Even so, masters migrating from Rugby, as well as past pupils exporting the school's football code to the universities and big commercial centres, made a good fist of spreading the gospel. Their evangelical work was so effective that in January 1871, 21 clubs attended a meeting called with the aim of forming a Rugby Football Union.

The historic meeting of rugby football's first recognised union (now known as England Rugby) was held at the Pall Mall Restaurant near London's Trafalgar Square. Eight of those founder clubs flourish to this day: Richmond, Blackheath, Harlequins, Civil Service, Guy's Hospital, Wellington College, King's College and St Paul's School. And a ninth club still in membership of the union would have been founder members had not the intrepid representatives of Wasps gone to the wrong London hostelry.

The urgent business of the fledgling union involved laying down a code of playing rules and in this respect the Rugby School Laws were more or less wholly adopted. The other pressing item for the new Union to consider was responding to the challenge issued by Scottish enthusiasts for a proposed Scotland–England international match to be played under Rugby rules.

FAMOUS FIRSTS

The youngest captain

England's first rugby international was a 20-a-side affair staged on Edinburgh's Raeburn Place on 27 March, 1871. They were beaten by a goal and a try to a try in the days before scoring by points, but their captain that day, Frederic Stokes of Blackheath, set a record which remarkably still holds. Stokes was 20 years and 258 days old when he led England and remains their youngest skipper in more than 145 years of Test rugby.

Leading the rose

Including Fred Stokes's pioneering act in 1871, 129 men have captained England in Test matches.

Name	First captaincy
F Stokes	27 Mar 1871
AStG Hamersley	23 Feb 1874
HA Lawrence	15 Feb 1875
F Luscombe	13 Dec 1875
E Kewley	5 Feb 1877
MW Marshall	11 Mar 1878
FR Adams	10 Mar 1879
L Stokes	2 Feb 1880
AN Hornby	6 Feb 1882
ET Gurdon	16 Dec 1882
CJB Marriott	2 Jan 1886
Alan Rotherham	8 Jan 1887
F Bonsor	16 Feb 1889
AE Stoddart	15 Feb 1890
JL Hickson	1 Mar 1890
FHR Alderson	3 Jan 1891
SMJ Woods	6 Feb 1892
RE Lockwood	6 Jan 1894
EW Taylor	17 Mar 1894
F Mitchell	14 Mar 1896
JF Byrne	5 Feb 1898
Arthur Rotherham	7 Jan 1899
RHB Cattell	6 Jan 1900
J Daniell	3 Feb 1900
JT Taylor	5 Jan 1901
WL Bunting	9 Feb 1901
H Alexander	11 Jan 1902
B Oughtred	10 Jan 1903
PD Kendall	21 Mar 1903
FM Stout	9 Jan 1904
VH Cartwright	2 Dec 1905
BA Hill	5 Jan 1907
J Green	9 Feb 1907
EW Roberts	16 Mar 1907
TS Kelly	1 Jan 1908
JGG Birkett	18 Jan 1908
CEL Hammond	8 Feb 1908
LAN Slocock	21 Mar 1908
GHD'O Lyon	9 Jan 1909

R Dibble	16 Jan 1909
AD Stoop	15 Jan 1910
ER Mobbs	3 Mar 1910
ALH Gotley	18 Mar 1911
NA Wodehouse	8 Apr 1912
RW Poulton	17 Jan 1914
JE Greenwood	17 Jan 1920
WJA Davies	15 Jan 1921
LG Brown	21 Jan 1922
WW Wakefield	19 Jan 1924
LJ Corbett	15 Jan 1927
R Cove-Smith	7 Jan 1928
HG Periton	16 Mar 1929
JS Tucker	22 Feb 1930
PD Howard	14 Feb 1931
CD Aarvold	21 Mar 1931
AL Novis	11 Feb 1933
BC Gadney	20 Jan 1934
DA Kendrew	19 Jan 1935
HG Owen-Smith	16 Jan 1937
P Cranmer	15 Jan 1938
H Toft	19 Mar 1938
J Mycock	18 Jan 1947
J Heaton	15 Mar 1947
EK Scott	3 Jan 1948
TA Kemp	17 Jan 1948
RHG Weighill	29 Mar 1948
NM Hall	15 Jan 1949
I Preece	26 Feb 1949
VG Roberts	20 Jan 1951
JM Kendall-Carpenter	10 Feb 1951
RV Stirling	16 Jan 1954
PD Young	26 Feb 1955
E Evans	21 Jan 1956
J Butterfield	17 Jan 1959
REG Jeeps	16 Jan 1960
RAW Sharp	19 Jan 1963
MP Weston	25 May 1963
JG Willcox	4 Jan 1964
CR Jacobs	22 Feb 1964
DG Perry	16 Jan 1965
DP Rogers	15 Jan 1966
PE Judd	11 Feb 1967
CW McFadyean	20 Jan 1968
JRH Greenwood	8 Feb 1969
R Hiller	20 Dec 1969
RB Taylor	18 Apr 1970
AL Bucknall	16 Jan 1971
JS Spencer	13 Feb 1971
PJ Dixon	26 Feb 1972
JV Pullin	3 Jun 1972
FE Cotton	18 Jan 1975
A Neary	15 Mar 1975
RM Uttley	15 Jan 1977
WB Beaumont	21 Jan 1978
SJ Smith	6 Feb 1982
JP Scott	5 Mar 1983
PJ Wheeler	19 Nov 1983
ND Melville	3 Nov 1984
PW Dodge	5 Jan 1985
RJ Hill	7 Feb 1987
ME Harrison	4 Apr 1987
J Orwin	23 Apr 1988
RM Harding	16 Jun 1988
WDC Carling	5 Nov 1988
CR Andrew	13 May 1989
PR de Glanville	23 Nov 1996
J Leonard	14 Dec 1996
LBN Dallaglio	15 Nov 1997
AJ Diprose	6 Jun 1998
MJS Dawson	20 Jun 1998

MO Johnson	14 Nov 1998
KPP Bracken	2 Jun 2001
NA Back	10 Nov 2001
PJ Vickery	22 Jun 2002
JP Wilkinson	9 Mar 2003
DE West	30 Aug 2003
JT Robinson	13 Nov 2004
ME Corry	12 Mar 2005
PH Sanderson	11 Jun 2006
MJ Catt	11 Mar 2007
SW Borthwick	10 Feb 2008
LW Moody	20 Mar 2010
NJ Easter	20 Nov 2010
MJ Tindall	4 Feb 2011
CDC Robshaw	4 Feb 2012
DM Hartley	23 Jun 2012
TA Wood	8 Jun 2013
GT Ford	25 Nov 2017
OA Farrell	10 Mar 2018

FAMOUS CHARACTERS

Rugby's first superstar

The first England rugby superstar was the Blackheath utility back Lennard Stokes. His international career began in 1875 when 20-a-side was the order of the day and by the time it ended six years later, teams of 15 had become firmly established. Stokes was a household name in either formation.

He used his acceleration and dodging powers to good effect in attack and was described as the longest drop-kicker in the Home Unions. Before scoring by points was introduced in the late 1880s, matches were decided on the majority of goals (converted tries or drop goals) scored. Stokes's match-winning potential was so great that in his 12-match career the national side lost only once and his 17 conversions stood for more than a century as the England career record until overtaken by Jonathan Webb.

Arthur Guillemard, an early historian of the English game, described Stokes as the outstanding back of the late nineteenth century: "It is not too much to say that at this post his equal, either in science or play, has never been seen from the date of the foundation of the Union," he wrote in 1892.

Stokes was appointed captain in 1880 and in his five matches in charge never led a losing side. "He enjoyed a wonderful hold on his men," continued Guillemard, who confirmed that his admirable play and a great reputation "naturally made him the idol of the ring".

FAMOUS FIRSTS

Test rugby's oldest cup

In 1878, when the mad dogs and Englishmen of the Calcutta RFC decided to stop going out in the midday sun to play their favourite game, the club decided to donate their assets to the RFU in London. Their surplus rupees were melted down and fashioned into a silver cup which the parent Union quickly christened "The Calcutta Cup", putting it up as the prize for the winners of the annual international fixture between England and Scotland. The first cup in Test rugby's trophy cabinet was contested in 1879 when England and Scotland played out a draw. England became the first team to lift the trophy the year later, beating the auld enemy by the convincing margin of two goals and three tries to a goal at Manchester and have held the trophy most often since.

STAT ZONE

England's highest scores

England's first international against Wales was an overwhelming win by seven goals, six tries and a dropped goal to nil at Richardson's Field in Blackheath in February 1881. There was no scoring by points in those days, but under modern values that result would translate into an 82–0 win – one of England's highest scores. Poor old Wales were deemed "not good enough" to warrant a match with the English the next season, but returned to the fixture lists the season after. England's ten highest scores since points were introduced in the late 1880s are:

Points	Opponents	Venue	Year
134	Romania	Twickenham	2001
111	Uruguay	Brisbane	2003
110	Netherlands	Huddersfield	1998
106	United States	Twickenham	1999
101	Tonga	Twickenham	1999
84	Georgia	Perth	2003
80	Italy	Twickenham	2001
70	Canada	Twickenham	2004
67	Italy	Twickenham	1999
67	Romania	Dunedin	2011

STAT ZONE

Most tries in a Test by a player

The first England player to cross for a brace of tries in a Test was William Hutchinson against Ireland at The Oval in 1877. It was the first 15-a-side international match and the wily little half-back clearly revelled in the extra space presented by having ten fewer men on the pitch.

In the corresponding fixture at Manchester in 1880, Henry Taylor, another half-back, scored England's first hat-trick of tries in a match, but in the inaugural Test against Wales the year later it was a forward, George Burton of Blackheath, who went one better scoring four times. Only 13 players have scored four or more tries for England in an international match:

Tries	Player	Opponents	Venue	Year
5	D Lambert	France	Richmond	1907
5	R Underwood	Fiji	Twickenham	1989
5	OJ Lewsey	Uruguay	Brisbane	2003
4	GW Burton	Wales	Blackheath	1881
4	A Hudson	France	Paris	1906
4	RW Poulton	France	Paris	1914
4	CJ Oti	Romania	Bucharest	1989

4	JC Guscott	Netherlands	Huddersfield	1998
4	NA Back	Netherlands	Huddersfield	1998
4	JC Guscott	United States	Twickenham	1999
4	JT Robinson	Romania	Twickenham	2001
4	NJ Easter	Wales	Twickenham	2007
4	CJ Ashton	Italy	Twickenham	2011

FAMOUS FIRSTS

Captain of rugby and cricket

1882 was a landmark year for Lancashire's AN "Monkey" Hornby. In the winter he led the England rugby team in their home internationals against Ireland in Dublin and Scotland at Manchester, and in the summer he was captain of the England cricket team that lost to Australia at The Oval – the famous Test defeat that gave rise to "The Ashes". England's only other dual rugby/cricket captain was AE "Drewy" Stoddart in the 1890s.

FAMOUS FIRSTS

The Triple Crown

In 1882–83, England beat Wales, Ireland and Scotland to become first holders of the then mythical Triple Crown, an expression devised by the press to denote the winning of all three matches against the other Home Union nations in the same season.

England have claimed the honour more often than any other country (25 times): 1883, 1884, 1892, 1913, 1914, 1921, 1923, 1924, 1928, 1934, 1937, 1954, 1957, 1960, 1980, 1991, 1992, 1995, 1996, 1997, 1998, 2002, 2003, 2014, 2016.

FAMOUS CHARACTERS

A successful captain

In every decade of the game's history there has been a forward who has stood out for his outstanding technique. The man who lorded it over his opponents in the early years of international rugby was Temple Gurdon.

He was well versed in the game at his school, Haileybury, before going up to Cambridge where he won his Blue. His college refused him leave to play for England in 1877, but he finally made his debut in the 1878 match against Scotland. "In every respect he was one of the best forwards that ever represented England," wrote a critic. Gurdon was a muscular forward who used his weight and strength effectively and was "usually to be found in the very heart of the scrummage". A deadly tackler, he was particularly adept at dribbling, a long-since defunct skill but an important attacking ploy in the early years of the game.

All told he played 16 times for England in a career that spanned nine seasons, a record that "evidenced his sterling worth" and which stood until the early twentieth century. Only once, moreover, was he on the losing side. He was also England's first successful captain, leading an unbeaten side on nine occasions including back to back Triple Crowns in 1883 and 1884, when "he set his players an excellent example".

FAMOUS FIRSTS

Played for two countries

Irish eyes weren't smiling when they lined up to face England at Manchester in 1892. Carded to play as a threequarter against them was one James Marsh, a Swinton doctor late of Edinburgh University and who had featured in Scotland's win against the Irish in Belfast barely three years earlier. The authorities must have taken a dim view of this because Marsh remains the only man who has played international rugby for two Home Unions. All told, however, four England international players have won dual rugby caps:

James Marsh	Scotland 1889 & England 1892
Frank Mellish	England 1920–21 & South Africa 1921–24
Barry Holmes	England 1949 & Argentina 1949
Jamie Salmon	New Zealand 1981 & England 1985–87

FAMOUS GAMES

The perfect Triple Crown

SCOTLAND 0, ENGLAND 5

Raeburn Place, Edinburgh, 5 March 1892

This was the last great shout for English rugby before the formation of the Northern Union (later the Rugby League) in 1895 deprived the national side of northern-bred raw-boned, tough-as-teak forwards and wily, elusive backs. The northern industrial towns supplied this side with 13 players, including its entire back division.

The class of '92 created a unique record for the International Championship, winning its three games without conceding a score to secure the perfect Triple Crown. Wales had been beaten 17–0 at Blackheath and Ireland 7–0 in Manchester before England travelled to Edinburgh seeking only their third win on Scottish soil in 21 years. A dour game was resolved just before half-time when William Bromet of Tadcaster went over beneath the posts and Heckmondwike's Dicky Lockwood converted for a 5–0 verdict. Twenty-one years would pass before the Triple Crown again sat on English heads.

FAMOUS FIRSTS

Family tries

Many sets of brothers have appeared for England, with several pairs even playing together in the national fifteen. But tries scored by brothers in the same match are as rare as hens' teeth. The first siblings to do so were Frank and Percy Stout of Gloucester in the 14–7 defeat of Wales at Rectory Field, Blackheath in April 1898. Nearly a century then passed before the Underwoods, Rory and Tony, repeated the feat. Cheered on by their enthusiastic mum, Mrs Underwood's boys crossed in the 26–12 win against Scotland at Twickenham in 1993 before doing so again in successive matches (against Romania and Canada) in the 1994 autumn internationals and once more, against Italy, during the 1995 Rugby World Cup pool stage.

14

FAMOUS CHARACTERS

The wily tactician

Adrian Stoop's name became synonymous with Twickenham when his club, Harlequins, decamped there when the new ground opened in 1909. Stoop, one of the great thinkers about back play, was already an England Test half-back at the time, but it was the exciting brand of open rugby that he brought to the club scene that transformed Harlequins into a leading London club and was the catalyst for a sea-change in the fortunes of the England side.

Stoop was installed as fly half and captain for England's first Twickenham international, against Wales in January 1910, and ignited the national side with the same attacking principles that had served Harlequins so well. The English hadn't beaten Wales for 12 years, but there was a sensational start to the match. Stoop caught the ball from the kick-off, but instead of taking the safe option and finding touch, immediately moved in-field to launch a surprise attack. Ten seconds later his brave move culminated in a try in the right-hand corner. What a start for spectators in the new ground's spacious enclosures.

England never looked back, went on to beat Wales 11–6 and finished the season as winners of the International

Championship for the first time since 1892. Stoop was hero-worshipped from Land's End to Hadrian's Wall. His unorthodoxy catapulted England out of a long period in rugby's doldrums and launched a long period of English dominance in the International Championship, though he was to retire from the international scene in 1912.

15

FAMOUS GAMES

A new dawn

ENGLAND 11, WALES 6

Twickenham, 15 January 1910

The RFU's new headquarters at Twickenham, having been purchased in 1907–08, was finally ready to stage major matches by the autumn of 1909, and what a start there was to this first international at the ground. Wales kicked off deep into the English 25 where Adrian Stoop, the home side's captain, immediately showed his predilection for attacking rugby by shunning the safe option of a kick to touch and instead launching an ambitious handling movement. Barely 10 seconds later his initiative culminated in a try in the right corner scored by Fred Chapman.

England had not beaten Wales since 1898 and the early 1900s had been a golden period of Welsh dominance of the Championship. Yet the new ground, the early score and Stoop's insouciance at fly half somehow galvanised the unfancied men in white to lift their game. Near half-time Dai Gent, who had once played in a Welsh trial before throwing in his lot with England, made a telling break from a loose scrum to send

Bert Solomon careering over for a try that placed the hosts comfortably ahead at 11–3 by the interval.

Stout English defence kept the Welsh in check for the rest of the match and they ran out deserved winners before eventually going on to claim their first Championship title since 1892.

FAMOUS CHARACTERS

The Welshman who became an England rugby great

English rugby was blessed by a guiding genius of a fly half behind the much vaunted packs of the pre- and post-Great War era. That the player was a Welshman by birth was largely ignored by the Fleet Street scribes of the time: W J A "Dave" Davies was the most-loved back in English rugby at the time.

He was born and raised in Pembrokeshire, that "Little England beyond Wales", but made his name in rugby circles through his stylish performances for Royal Navy and United Services sides. While Davies had all the running and passing skills of his distinguished predecessor, Adrian Stoop, it was the long-raking tactical kicking and accurate left-footed drop-kicking that set him apart as England's finest exponent of the fly half's arts.

All told, Davies featured in four England Grand Slam sides, captaining the successful teams of 1921 and 1923, and was never on a losing England side in the International Championship. England's only defeats during his career span were against Wales in 1920 and 1922 when Davies was absent through injury.

The key to the national side's post-war successes was Davies's half back partnership with his naval colleague, Cecil Kershaw. While stationed in Portsmouth, the two regularly met after their daily naval duties were finished and forged a strong understanding of each others' play through countless hours of training together.

FAMOUS FIRSTS

The Grand Slam

The Holy Grail of the Five (and later Six) Nations tournament is winning the Grand Slam – defeating all opponents in the tournament in the same season. Although the expression was not applied in a rugby context until 1957, England's first clean sweep of the other European nations was in 1913. They have now won the accolade more times than any other nation (13 times): 1913, 1914, 1921, 1923, 1924, 1928, 1957, 1980, 1991, 1992, 1995, 2003, 2016.

GRAND SLAM
RBS 6 NATIONS CHAMPIONS 2016
RUGBY'S GREATEST CHAMPIONSHIP
RBS
6
NATIONS

RBS 6 NATIONS®

18

FAMOUS CHARACTERS

The father of English flankers

The player credited with developing the distinctive feature of the modern flanker role in English rugby was the Blackheath forward, Charles Pillman. In the days before forward specialisation became the norm in English rugby, he packed on the side of the scrum where he could quickly detach himself and used his pace to attack and disrupt the opposing fly half. So successful was his technique that the selectors nominated him to perform his specialist duties against Wales in 1910 in the first international staged at Twickenham.

Aged 20, Pillman enjoyed a dream debut playing an important part in a momentous victory and it was widely acknowledged that the youngster's effective winging operations helped England to go on and win the first Five Nations title. Pillman showed a remarkable ability to read a game, often lending a helping hand to his colleagues in attacks and was quite simply the 'find' of the year.

"Cherry" Pillman was an automatic choice for the England packs that laid the foundations of four Championship titles including two Grand Slams in the seasons leading up to the outbreak of the 1914–18 war and his legacy was that by the

1920s every international side adopted England's example and fielded a winger in the Pillman mould.

19

FAMOUS CHARACTERS

The deadliest finisher

The smallest international player of his generation, "Kid" Lowe occupied the England right-wing berth for ten years before retiring in 1923. He never missed an international and but for the Great War that robbed him of five seasons of international rugby would almost certainly have played nearly 50 Tests – a phenomenal record considering that the only regular opponents of the day were Scotland, Ireland, Wales and France.

Lowe was the deadliest finisher England had before the advent of Rory Underwood. His eight tries in the 1914 campaign have never been surpassed in the history of the International Championship (despite the fact that there were only four opponents in those days) while his 18 tries stood as the overall England career record for 67 years and helped the side to four Grand Slams. Against Ireland at Twickenham in 1921, moreover, he showed that he wasn't just a runner in of tries. Finding his way to the try-line blocked, he swivelled towards the posts and promptly dropped a left-footed goal that was described as the "most delightful" score of the match.

20

BLOOD ON THE ROSE

The First World War

England's rugby players volunteered with alacrity to serve King and country when war broke out in August 1914. Few were quicker off the mark than England's Jack King, who enlisted with the Yorkshire Hussars on 6th August 1914 and left his farm three days later to begin his service career.

Originally told he was too short (one inch below the regulation 5ft 6in required), Jack stood his ground and told the recruiting officer in no uncertain terms that he would "simply stick here until you do take me in". Military records show that he officially became a trooper on 12th August, undertook his training in Hitchin, Herts and went off to France.

International players featured prominently among the war-dead. Of the 30 who had taken part in the Triple Crown match between Scotland and England in 1914, 11 gave their lives and, all told, 27 former England caps were killed in action, lost or died of wounds. Among them were King (killed in action on the same day and at the same place as Lancelot Slocock), Ronald Poulton-Palmer, John Raphael and Robert Pillman, younger brother of "Cherry" Pillman whose place he'd taken in the last England side to play before war.

Arthur Harrison, a forward who had played against Ireland and France in the 1914 Grand Slam campaign, headed the list of casualties decorated in battle. He was posthumously awarded the V.C. for his gallantry during the blockade of Zeebrugge.

ROLL OF HONOUR 1914–1918

WILSON, Charles Edward (England 1898) KIA River Aisne, September 17, 1914.

WATSON, James Henry Digby (England 1914) Lost at sea when HMS *Hawke* was torpedoed on October 15, 1914.

OAKELEY, Francis Eckley (England 1913–14) Lost at sea between November 25 & December 1, 1914.

KENDALL, Percy Dale (England 1901–03). KIA Ypres, January 25, 1915.

LAGDEN, Ronald Owen (England 1911) KIA St Eloi, March 3, 1915.

TODD, Alexander Findlater (England 1900) Died of wounds, Ypres, April 21, 1915.

POULTON-PALMER, Ronald William (England 1909–14) KIA Ploegsteert Wood, May 5, 1915.

BERRY, Henry (England 1910) KIA Festubert, France, May 9, 1915.

NANSON, William Moore Bell (England 1907) KIA Gallipoli, June 4, 1915.

TARR, Francis Nathaniel (Frank) (England 1909–13). KIA Ypres, July 18, 1915.

DINGLE, Arthur James (England 1913–14) KIA Gallipoli, August 22, 1915.

LAMBERT, Douglas (England 1907–11) KIA Loos, October 13, 1915. Loos Memorial.

ALEXANDER, Harry (England 1900–02) KIA Hulluch, northern France, October 17, 1915.

PILLMAN, Robert Lawrence (England 1914) Died of wounds Armentières, July 9, 1916.

HAIGH, Leonard (England 1910–11) Died in training at Woolwich ,August 6, 1916.

SLOCOCK, Lancelot Andrew Noel (England 1907–08) Died of wounds Guillemont, August 9, 1916.

KING, John Abbott (England 1911–13) KIA Guillemont, August 9, 1916.

INGLIS, Rupert Edward (England 1886) Killed rescuing wounded in Ginchy, France, September 18, 1916.

MAYNARD, Alfred Frederick (England 1914) KIA Beaumont Hamel, November 13, 1916.

DOBBS, George Eric Burroughs (England 1906) KIA Poperinghe, June 17, 1917.

WILSON, Arthur James (England 1909) KIA Flanders, July 31, 1917.

MOBBS, Edgar Roberts (England 1909–10) KIA, Zillebeke, July 31, 1917.

HODGES, Harold Augustus (England 1906) KIA, near Mons, March 24, 1918.

HANDS, Reginald Harold Myburgh (England 1910) Died of wounds at Boulogne, April 20, 1918.

HARRISON, Arthur Leyland (England 1914) KIA Zeebrugge, April 23, 1918.

RAPHAEL, John Edward (England 1902–06) Died of wounds Remy (Messines), June 11, 1917.

SCHWARZ, Reginald Oscar (England 1899–1901) Died of influenza on active service, Etaples, November 18, 1918.

21

FAMOUS CHARACTERS

The moderniser

The dominant personality of English rugby during its Golden Era in the early 1920s was Wavell Wakefield. Although a forward of immense physical strength and all-round ability, it was his tremendous pace that brought him six Test tries during a career that produced the then record number of caps for an Englishman (31).

But it was Wakefield's ability to think about the game that underpinned England's three Grand Slam wins of 1921, 1923 and 1924. He brought his analytical mind to the mechanics of forward play at a time when "first up, first down" was the accepted system for scrummaging in Britain. This meant that front rows of scrums were invariably formed by the first three forwards arriving at the scene. But as England's pack leader in 1923, Wakefield set about developing fixed positions for his forwards at scrums, with specialist props, hookers, second rows and loose forwards performing their respective duties at the set piece.

His pack, which showed only one change during the season, played with such cohesion that England swept all before them on the way to a Grand Slam, and the next season, as captain of a younger and much changed fifteen, Wakefield's methods again brought dividends as England won back-to-back Grand Slams for the first time. He captained England for two more seasons before winning his final cap in Paris in 1927. After retirement, Wakefield went on to make a valued contribution to the RFU as a thoughtful administrator.

GOLDEN OLDIES

England's oldest player

When full-back Fred Gilbert of the Devonport Services took the field at Twickenham for the match against Wales in 1923 he set the record as England's oldest debutant. His selection was criticised at the time, but he proved a valuable asset to his side in a 7–3 win and retained his place for the next match, against Ireland at Welford Road where he set the record as the oldest player to represent England – at 39 years 42 days. He was injured playing for the Royal Navy at Twickenham a couple of weeks later, left the field at half-time and never played any first-class rugby again.

STAT ZONE

Fastest try

England's opening fixture of their 1923 Grand Slam campaign saw Wales visit Twickenham on a windy day in London. From the kick-off, the Leicester flanker Leo Price scored a try before a Welshman had even touched the ball in a phase that was clocked at ten seconds. England's captain, "Dave" Davies, takes up the story.

"Wales were beaten after one of the most sensational starts ever seen at a rugby football match. England kicked off against a small gale blowing straight down the ground. The ball was blown into the hands of H.L. Price, who was following up. He endeavoured to drop a goal. The ball fell short on to the goal-line. Price caught it and fell over the line for a try before the unhappy Welsh team realised the game had commenced."

FAMOUS FIRSTS

The personal clean sweep

When England won the 1924 Grand Slam their new right-wing, Carston Catcheside, set an unusual record. He scored two tries in the season-opener against Wales in Swansea, grabbed another brace when England visited Belfast to meet Ireland three weeks later, and was again on the score-sheet crossing for tries in the Twickenham matches against France and Scotland that sealed the Slam for his country and a unique personal clean sweep. In doing so he became the first player to score tries in every match of a country's Five Nations campaign.

His feat was subsequently equalled by players representing other sides in Five and Six Nations tournaments, but he is the only Englishman to achieve this unusual record to date.

The penny pincher

The same Carston Catcheside even managed to score one over the ever-parsimonious RFU in 1924. The cheeky wing from Percy Park in Northumberland had put in a rounded expenses claim for £4 following a Twickenham trial match before winning selection for the international team, only to find the claim reduced to the correct rail-fare, £3-19-11. After winning the Grand Slam against Scotland at Twickenham in March, he again submitted for £4 – this time successfully securing a refund for spending a penny at Kings Cross:

To third-class railway fare Newcastle to Kings Cross, return – £3-19s-11d
To use of toilet at Kings Cross Station – 1d
TOTAL CLAIM – £4-0s-0d

26

FAMOUS FIRSTS

Twickenham calling

The first live rugby match broadcast was the England–Wales clash from Twickenham in January 1927, soon after the BBC became a national corporation. Sports broadcasting had been successfully pioneered in the United States and the fledgling Beeb, anxious to broaden its scope, decided that a rugby match would be the setting for British sport's first running commentary. Teddy Wakelam, a former Harlequins player, was given the mic and a blind man as companion so that he could broadcast as if describing the run of play to him. Wakelam later recalled: "Our perch was a somewhat rickety-looking hut mounted on a platform at the end of the West Stand in the south-west corner of the ground and the only advice offered was in the form of a notice printed in large red letters reading, 'DON'T SWEAR.'"

27

FAMOUS FIRSTS

The flying hooker

In 1930, England's Sam Tucker, who had been first capped in 1922, was summoned to make an extraordinary last-minute journey from Bristol to Cardiff by plane and lorry to replace a late-hour withdrawal. The first man to fly to an international, he inspired a rare win at the Arms Park and ended the year as captain for a scoreless draw with Scotland that earned England an unexpected Five Nations title.

Sam Tucker (England 1922–1931, 27 caps): On Saturday morning, January 18, 1930, after having a glass of beer in a Bristol hostelry, I returned to my office to clear up a few things when, at 12.25 pm, the telephone rang. It was Engineer-Commander S.F. Coopper, secretary of the R.F.U.

He said: "We want you to come over to Cardiff to play against Wales today!" I said it was impossible, as the last train had gone, but then selector John Daniell came on the 'phone to say: "You must be there, even if you come by 'plane."

The only thing I could do was call Filton Aerodrome, as there was no landing strip in either Bristol or Cardiff in those days. A lady answered the 'phone [and] said there was a certain "Captain Somebody" in the air and she would try to get a signal to him to come down.

By a stroke of luck she succeeded and the "Captain" phoned me at 1pm. He said if I went out at once he would get me to Cardiff between 2pm and 2.15. I was at Filton [by 1.50] and there was the pilot with a small bi-plane two-seater already revved up. He stuck a helmet on my head, strapped me in and, with a roar, we were off.

We arrived over Cardiff at roughly 2pm and after circling around he thought he could land in a field. I thanked him, raced over the field on to the road and waved the driver [of a coal lorry] down [who] did the journey [to the Arms Park] in under ten minutes. We won 11–3, and somehow I became a bit of a hero.

28

FAMOUS GAMES

Beating the All Blacks

ENGLAND 13, NEW ZEALAND 0

Twickenham, 4 January 1936

"A remarkable match, with a still more remarkable result," reported the man from *The Times* after the All Blacks were beaten on English soil for the first time. The Prince of Wales and 70,000 spectators witnessed a victory that was ascribed to the solidity of a heavy pack ably assisted by a close-marking midfield back division who showed a fine turn of speed in attack. But the match will forever be inextricably linked to the name of a Russian prince, Alex Obolensky, whose two first-half tries put paid to any hopes New Zealand might have entertained regarding victory.

First he turned his marker inside out and outstripped the defence with a straight run of 40 metres along the right wing to open the scoring after 20 minutes. Then, on the stroke of half-time, his turn of pace took him on an unorthodox cross-field run. He not only rounded the All Blacks through sheer pace, but in so doing lost his own support players. Having veered in sharply from the right, he finished off by turning in from the left wing to score a wonderful solo try that was greeted with a loud and enthusiastic fervour rarely witnessed at Twickenham. Peter Cranmer dropped a goal from 30 yards early in the second half and New Zealand's rout was complete when he sent Hal Sever sprinting 35 yards for the final try.

FAMOUS CHARACTERS

The Russian prince

He set no special records for England. Indeed, he only appeared four times in the white jersey with the red rose. Yet his deeds against the All Blacks on a cold January day in 1936 made him a household name. The player was Alex Obolensky, an exiled Russian prince who was studying at Oxford University.

The two tries he scored against New Zealand on his debut made "Obo" an overnight star. He was an electric runner who used his acceleration to deceive opponents, but it was his love of the unorthodox that marked him out from the run-of-the-mill wings of the day. His second try against the All Blacks, in particular, was a classic. Receiving the ball far out on his right wing, he ghosted through the New Zealand defence to finish off scoring wide out to the left of the posts.

Captured on film by Movietone News, it was probably the first famous try to be seen by a broad audience. Vivian Jenkins, the Welsh full back at the time, recalled that he spent several hours at an Oxford Street cinema specially to see the two or three minutes of newsreel between feature films. (Wales were due to play England a fortnight later.) Obolensky's fame has never waned, though he never again set the rugby world alight. He kept his place for the three Championship matches in 1936 – a scoreless draw with Wales, defeat in Dublin and a one-point win against Scotland back at Twickenham. But he never again appeared in a cap match and was killed tragically young in a flying accident early in World War II.

FAMOUS FIRSTS

The unique Triple Crown captain

When England won the Triple Crown in 1937, their captain was the St Mary's Hospital medical student "Tuppy" Owen-Smith. It's true that no side ever won a Triple Crown by a narrower set of margins – 4–3 against Wales, 9–8 over Ireland and 6–3 against Scotland in England's first victory at Murrayfield – but the full back's main claim to sporting fame was that he had scored a Test cricket century for his native South Africa against his adopted country during the Headingley Test of 1929. While several overseas Test cricketers have also played rugby for England, Owen-Smith's dual feats are unique.

FAMOUS FIRSTS

Televised rugby

There couldn't have been a better match selected for the first live telecast of an international. In March 1938, the expanding BBC television department chose to broadcast live the Calcutta Cup match on a sunny March day that had a touch of spring in the air. Both England and Scotland could still win the Championship, but in a match of constant thrills at Twickenham there was never more than a converted try between the sides. Wilson Shaw scuttled away for a famous late score, regaining the Triple Crown for Scotland, but the audience was limited because the BBC's television reach extended no further than a twenty-mile radius of broadcasting house in London. In November 1967, England were again television pioneers when their autumn Test defeat by the All Blacks, also at Twickenham, was the first match shown live in colour.

32

BLOOD ON THE ROSE

The Second World War

War against Hitler was different from conflict with the Kaiser. Once again there was no shortage of volunteers from the rugby brotherhood, but the long period of "Phoney War" enabled rugby, through the organising abilities of the services, to continue more extensively than during World War One. By the end of September 1939, an informal season was announced and many fixtures took place between service units, students and even the clubs continued by recruiting from servicemen stationed nearby.

This war was also remarkable for an armistice between Union and League. Animosities were suspended for players in the services and many of the 13-a-side's box-office boys – ex-Welsh union star Willie Davies and his famous League compatriot Gus Risman to name two – turned out many times in top-draw Union matches. There was even a famous cross-code challenge at Bradford's Odsal Stadium, League's equivalent of Twickenham (which had been requisitioned and suffered damage from a V-bomb blast in July 1944).

There were mercifully fewer killed in action this time, though among the 14 English casualties were two names from the win against the All Blacks: Ron Gerrard and Alex Obolensky, who was the first England cap to perish when his aircraft crashed during a training flight in Suffolk in 1940 – another story in the continuing legend of the fabled Prince.

ROLL OF HONOUR 1939–1945

OBOLENSKY, Alexander (England 1936) Killed flying, Martlesham, Suffolk, March 29, 1940.

COOKE, Paul (England 1939) KIA Calais, May 1, 1940.

BLACK, Brian Henry (England 1930–1933) Killed Chilmark, Wilts, July 29, 1940.

PARSONS, Ernest Ian (England 1939) Killed flying over the Channel, August 13/14, 1940.

TEDEN, Derek Edmund (England 1939) Killed flying over North Sea, October 15, 1940.

REW, Henry (England 1929–1934) Died of wounds, Western Desert, December 11, 1940.

LUDDINGTON, William George Ernest (England 1923–1926) Lost at sea, off Sicily, January 10, 1941.

TANNER, Christopher Champain (England 1930–1932) Died rescuing survivors of a sinking off Crete, May 23, 1941.

WODEHOUSE, Norman Atherton (England 1910–1913) Lost at sea off West Africa, July 4, 1941.

DAVIES, Vivian Gordon (England 1922–1925) KIA London, December 23, 1941.

FREAKES, Hubert Dainton (England 1938–1939) Killed Honeybourne, Worcs, March 10, 1942.

BOOTH, Lewis Alfred (England 1933–1935) KIA flying, June 25, 1942.

GERRARD, Ronald Anderson (England 1932–1936) KIA Western Desert, January 22, 1943.

MARSHALL, Robert Mackenzie (England 1938–1939) Lost at sea off Skagerrak, May 12, 1945 at the end of the war.

WARTIME RUGBY

The Second World War

Red Cross Internationals between England and Wales were staged in 1939–40 before giving way to a long series of Services Internationals in the middle years of the war. Then, when peace was declared, a season of "Victory Internationals" followed in 1945–46. None of the matches warranted full caps because, it was argued, the best players weren't always available owing to the exigencies of service. Even during the post-war "Victory" series many were still stationed abroad awaiting demob. That was a pity for Ray Longland, the nuggety, barrel-chested Northampton prop who had been a winner against the All Blacks and an England regular since 1932. He played 16 wartime "internationals" up to 1945 making him the true forerunner in skill, character and endurance of the modern era's Jason Leonard – even though the official record doesn't show it.

Wartime and Victory International results:

Red Cross Internationals (1939–40)

9 Mar 1940	Won	18–9 v Wales (Cardiff)
13 Apr 1940	Won	17–3 v Wales (Gloucester)

Services Internationals (1942–45)

7 Mar 1942	Lost	12–17 v Wales (Swansea)
21 Mar 1942	Lost	6–21 v Scotland (Inverleith)
28 Mar 1942	Lost	3–9 v Wales (Gloucester)
11 Apr 1942	Lost	5–8 v Scotland (Wembley)
7 Nov 1942	Lost	7–11 v Wales (Swansea)
27 Feb 1943	Won	29–6 v Scotland (Inverleith)
20 Mar 1943	Lost	7–34 v Wales (Gloucester)
10 Apr 1943	Won	24–19 v Scotland (Leicester)
20 Nov 1943	Lost	9–11 v Wales (Swansea)
26 Feb 1944	Won	23–13 v Scotland (Murrayfield)
18 Mar 1944	Won	27–15 v Scotland (Leicester)
8 Apr 1944	Won	20–8 v Wales (Gloucester)
25 Nov 1944	Lost	11–28 v Wales (Swansea)
24 Feb 1945	Lost	11–18 v Scotland (Leicester)
17 Mar 1945	Won	16–5 v Scotland (Murrayfield)
7 Apr 1945	Lost	9–24 v Wales (Gloucester)

Victory Internationals (1945–46)

24 Nov 1945	Lost	3–18 v NZ Kiwis (Twickenham)
19 Jan 1946	Won	25–13 v Wales (Cardiff)
9 Feb 1946	Won	14–6 v Ireland (Dublin)
23 Feb 1946	Lost	0–3 v Wales (Twickenham)
16 Mar 1946	Won	12–8 v Scotland (Twickenham)
13 Apr 1946	Lost	0–27 v Scotland (Murrayfield)

BEFORE AND AFTER

Capped either side of the war

Three England players' cap careers spanned the long break caused by the Second World War:

PLAYER	CAREER SPAN
Jack Heaton	(1935–1947)
Tommy Kemp	(1937–1948)
Dick Guest	(1939–1949)

35

FAMOUS CHARACTERS

The round-the-corner kicker

His father was Irish and his mother Scottish, but Norman "Nim" Hall was the backbone of the England side for the best part of nine seasons when official internationals restarted after World War II. He entered first-class rugby as a young 18-year-old medical student in 1943 and was a prominent member of the St Mary's Hospital back division that enjoyed a long period of success during the restricted programme of club matches staged in the war seasons. Later, though still in his playing prime, Hall was called "Rugby's Methuselah" by critics who joked that he had been in the game for so long.

His exploits with the hospital side made him an automatic choice for England when organised Tests restarted in 1947. On his debut he helped sink Wales at Cardiff with one of his trademark dropped goals and he was the side's linchpin at fly half in a season that brought a share of the International Championship. Only 23 when he was made captain in 1949, his appearances over the next three seasons were restricted through injury, but he returned as captain in 1952 when England were runners-up in the Championship. As a converted full back, he led an enterprising side to the title in 1953 when his style and certainty as a then unprecedented round-the-corner place-kicker yielded the two important penalties that earned England a lucky draw in Dublin. Still under 30, he last appeared in 1955 winding up a career in which he had led his country 13 times to match Wavell Wakefield's old record.

GOLDEN OLDIES

The longest-lived member of the brotherhood

Jason Leonard, the great England prop whose long Test career spanned the amateur and professional eras, refers to those who have worn the red rose of England as "The Brotherhood". One of his predecessors in the England front-row, Harry Walker of Coventry, became England's longest-lived international player. Harry was an ever-present in the first nine England post-war teams. First capped against Wales at Cardiff in 1947 when he was a month short of his 32nd birthday, he remained sprightly into his old age, vividly recalling in 2014 for Stephen Jones and Nick Cain, the authors of *Behind the Rose*: "Eric Evans was brought in [against Australia in 1948] and said that he wanted to play tight-head. I told him I couldn't care less, so he started practising at tight-head. Then on to the park came the selectors and they asked me what I thought I was doing. I told them Eric wanted to play tight-head. They looked at Eric and told him that if he wouldn't come round to the loose-head side he could bugger off. Eric was a good hooker, he was never a prop."

STAT ZONE

The most inexperienced fifteen

When England took the field at Cardiff Arms Park in January 1947 for their first official cap match for eight years, there were 14 new caps in the side, including skipper Joe Mycock. That remains the England record for 15-a-side Tests. Only once, in the days before 1877 when teams were 20-a-side, had England fielded more new caps in an international starting fifteen, and that was in their first match.

NEW CAPS	OPPONENTS	VENUE	YEAR
20	Scotland	Edinburgh	1871
14	Scotland	London	1872
14	Wales	Cardiff	1947
13	Wales	Gloucester	1900

38

FAMOUS CHARACTERS

Fitness fanatic 1

He won his first cap as a prop, but it was as an expert hooker and inspirational captain that Eric Evans is chiefly remembered. His England appearances were few and far between in the three years that followed his 1948 debut, but at length he resolved to improve his physical fitness and took himself off to train with the legendary Manchester United Busby Babes. He raised his stamina to levels that would not look out of place in today's demanding professional era, at the same time honing his skills to become one of the best technical exponents of the hooker's arts.

After featuring in the Championship-winning side of 1953 and the Triple Crown success of 1954, he was recalled in 1956 as captain of an inexperienced team. Aged nearly 35, Evans relished the challenge of leading a young England fifteen that contained ten new caps and, although seeing the team narrowly beaten by Wales in the opening game, the selectors wisely resisted the temptation to make wholesale changes. Evans's enthusiasm rubbed off on a side that developed the greatest admiration for their captain. He set them high standards with his own personal fitness and forged the entire team into such a formidable unit that over the three seasons he was in charge England lost only once more. They achieved the Grand Slam for the first time for 29 years in 1957 and won the Championship title again in 1958 before Evans retired to become ranked among the top England captains of all time.

FAMOUS FIRST AND LAST

The changing value of the dropped goal

For more than 75 years the dropped goal was the king of rugby's scoring actions. Before points were introduced, a majority of goals determined results, so a single dropped goal was worth more than any number of unconverted tries. Then, when points were introduced to decide international matches in the late 1880s, the dropped goal carried a higher value than a try. It was not until 1948 that the International Board reduced the value of the scoring action from four points to three, putting the drop on a par with a try. England's last four-point dropped goal was scored by "Nim" Hall, against Scotland in a blizzard at Twickenham in 1947. The same player also scored their first three-pointer: against Wales at Cardiff in 1949.

40

FAMOUS CHARACTERS

Fitness fanatic 2

"At centre . . . England gained considerably in penetrative power through the introduction of Butterfield." That was the considered opinion of *The Times*'s rugby correspondent, Owen Owen, after Jeff Butterfield's debut in 1953. Six years and 27 more successive matches that included four Championship titles, two Triple Crowns and a Grand Slam would not alter that judgement. Like his England skipper Eric Evans, the hallmark of Butterfield's approach to Test rugby was personal fitness. He set high standards working to improve his strength and speed, but while the hours of training gave him the stamina to tackle opponents to a standstill, the striking feature of his play was his handling. A natural sportsman, he had the dexterity to take and give a pass in one smooth action and the vision to straighten a threequarter attack. Often his carefully weighted transfers created tries for his wings, though he was a deadly finisher in his own right, well able to score tries from his searing breaks.

Butterfield's vision was a key element in England's successful years between 1953 and 1958. Just consider the statistics: during his career 21 of the 43 tries England scored were credited to threequarters and England kept 15 clean sheets. International rugby in the fifties was not the fast-flowing spectacle that it is today, maybe. Yet Jeff Butterfield's skills would have shone like a beacon in any era.

FAMOUS LASTS

The last dual international

Now that rugby and cricket are year-round international sports, there will never be another double-international at Test level in these games. The last of the 13 England rugby caps honoured as Test cricketers was MJK (Mike) Smith whose sole appearance for the rose was as fly half partnering Dickie Jeeps against Wales at Twickenham in 1956. He was capped from Hinckley having won rugby Blues at Oxford in 1954 and 1955. His England Test cricket debut came two years later against New Zealand.

The England rugby caps who played Test cricket were:

PLAYER	RUGBY CAPS	CRICKET CAPS
WH Milton	2, 1874–75	3 for South Africa, 1889–92
AN Hornby	9, 1877–82	3 for England, 1879–84
GF Vernon	5, 1878–81	1 for England, 1882–83
AE Stoddart	10, 1885–93	16 for England, 1888–98
SMJ Woods	13, 1890–95	3 for Australia, 1888 & 3 for England, 1896
F Mitchell	6, 1895–96	2 for England, 1899 & 3 for SA, 1912
RO Schwarz	3, 1899–1901	20 for South Africa, 1906–12
RH Spooner	1, 1903	10 for England, 1905–12
RHM Hands	2, 1910	1 for South Africa, 1914
HG Owen-Smith	10, 1934–37	5 for South Africa, 1929
MP Donnelly	1, 1947	7 for New Zealand, 1937–49
CB van Ryneveld	4, 1949	19 for South Africa, 1951–58
MJK Smith	1, 1956	50 for England, 1958–72

FAMOUS CHARACTERS

The flanker who always got his man

Peter Robbins was the out-and-out open-side flanker who tackled like a thunderbolt and covered the ground with lightning pace in the packs that dominated the Five Nations in 1957, 1958 and 1960. An Oxford graduate, he lined up under Eric Evans among the ten new caps who faced Wales in 1956 and quickly gained a reputation as a canny flanker who used his sense of anticipation to nail opposing fly halves. Robbins, like the Mounties, always got his man. Teams coming up against him in a dozen Five Nations matches between 1956 and 1960 – he was controversially dropped in 1959 because he didn't push in the scrums, the critics said – managed only three tries. That statistic underlines the effective defensive work of the team and of the flankers in particular.

Many critics canvassed him as an England captain. Robbins had lifted an ordinary Oxford side into a fit and effective unit that outwitted Cambridge to win the 1957 Varsity match. But the call never came. He was succeeded on the open side by Budge Rogers in 1961 and despite subsequently switching to the blind side for his club, his tally of appearances ended at 19 after a brief recall in 1962. He should, many felt, have won twice as many caps.

43

FAMOUS GAMES

Grand Slam in front of the Royal Family

ENGLAND 16, SCOTLAND 3

Twickenham, 16 March 1957

Under Eric Evans, England enjoyed three successful seasons between 1956 and 1958, losing only two of their 13 internationals. Arguably their finest hour came in this three-try win set before the Queen, Prince Philip and Prince Charles. The term "Grand Slam", moreover, referring to an invincible Championship season, appeared for the first time in the press reviews that season. Forward dominance and a back division that snaffled up its chances helped England to their first Grand Slam for 29 years. It was also to be their last for another 22 seasons.

In a tentative start, England reacted nervously to the few chances that came their way and an expectant crowd began to wonder if the enormity of the occasion would prove too burdensome for a young side. But doubts began to evaporate before half-time when Jeff Butterfield, at his creative best, carved out an opening for his co-centre Phil Davies to cross at the corner to give England a 6–3 lead. It was a try that whetted appetites for the feast that was to follow late in the match, when the Scottish defence finally crumbled. Flanker Reg Higgins, who played a blinder, burst away to send Peter Thompson over and scored the last try himself after unrelenting English pressure near the Scotland line.

FAMOUS CHARACTERS

The master of the unexpected

Peter Jackson was the most unorthodox wing to play for England since Alex Obolensky. Like the Russian prince, beating players came instinctively to him. He could tear defences apart with a lethal cut-in that wrong-footed opponents, but the criticism levelled at him was that his unpredictability too often bamboozled his own colleagues.

He entered Eric Evans's England team alongside nine other new caps against Wales in 1956 and in the following match against Ireland scored the first of his six international tries. Jackson was an inspiration to the 1957 Grand Slam side. His reputation as a tricky runner now well established, he had opponents jittery at the prospect of marking him. Against Wales his vis-à-vis fell off-side at a set piece to present England with a penalty goal near the posts which proved to be the only score of the game. Jackson scored the only try of the match in Dublin and crossed for two beauties in the 9–5 defeat of France at Twickenham.

He thrilled Twickenham again in 1958, scoring the winning try against Australia with a mesmerising solo effort in the dying moments and he was a success in New Zealand with the 1959 Lions before mysteriously losing form and favour. Jackson made only one more appearance in the next three seasons before returning, aged 32, to play in the side that won the Championship title outright in 1963. His pace against Wales created the spectacular match-turning try by Malcolm Phillips in what was destined to be England's last Cardiff triumph for 28 years.

45

FAMOUS GAMES

Last-gasp finish

ENGLAND 9, AUSTRALIA 6

Twickenham, 1 February 1958

There have been more polished performances and more important victories. But for sheer guts or a more desperate grandstand finish there can be few among England's 700-plus Tests to match this winter win against Bob Davidson's Fourth Wallabies.

Several England players were concussed or injured during the course of a rough game and, 11 years before the use of substitutes was finally sanctioned at Test level, Eric Evans's team had to play for nearly an hour with only 14 men. Phil Horrocks-Taylor had been injured early on, necessitating a wholesale rearrangement of the back division.

In the dying minutes the scores were tied at 6–6 when, with Jeff Butterfield also lying injured, Eric Evans exhorted his men to raise one last effort to win the game. The player who turned the match with a sublime do-or-die run was right wing, Peter Jackson. Receiving the ball on the Aussie 25, he deceived his opponent with a swerve and hand-off and ran clear to turn the full back inside out with a wonderful feint and check. Then he raced on to the line, winning a foot race with Rod Phelps to lunge over at the corner flag for the score that gave England their first success against a touring side for 22 years.

STAT ZONE

The longest wait

England's supporters had to endure 466 minutes without seeing their favourites score a try in the late fifties. Peter Jackson's intercept and 75-yard run for a try against France just before half-time on 1st March 1958 was their last until Jim Roberts went over in the 25th minute of the match against Wales at Twickenham on 16th January 1960. In between, England completed five international matches without scoring a try.

The second longest wait was 438 minutes, between Tony Underwood's score against Scotland at Twickenham in 1993 and his brother Rory's try against Wales, also at Twickenham, in 1994. Matches were played more frequently by then, but the try drought still included five whole matches.

47

FAMOUS CHARACTERS

The India-rubber man

Competitive, courageous, stroppy even; Dickie Jeeps was the scrum half England's forwards loved to have aboard. He never hesitated to stick his oar into the tide of players bearing down on him and made life hell for his opponents. No matter how severe a battering he took, he always bounced back ready for more, a quality that earned him the description, India-rubber man.

One of the rare band of international players to play in a Test for the Lions (1955) before winning his cap for his country (1956), he came to the fore in the 1957 Grand Slam season. His immaculate service from the scrum was the stuff of dreams for an attacking fly half like his partner that year, Ricky Bartlett. Jeeps was a fast and accurate passer during a period when tight marking was the norm in international matches. Jeeps's short passes just in front of his partner tempted the fly half to take the ball on the run. That way Bartlett was able to engage his talented threequarter line to unlock defences.

Jeeps's finest hour came in the match against Scotland when England wrapped up all the silverware the Championship had to offer. He bobbed about the field like a cork floating in choppy waters, at first bearing the brunt of Scottish back-row attacks before settling to take a hand in all three of England's tries. In 1960 he was England's new captain for a Championship season which brought another Triple Crown and a share with France of the Five Nations title. Jeeps was an inspiring captain who tackled resolutely and shrugged off the constant batterings he received, particularly from a rampaging French pack in Paris. He retired in 1962 after leading the side in 13 successive Tests. At the time he was England's most-capped scrum half.

48

FAMOUS GAMES

Sharp the guiding genius

ENGLAND 14, WALES 6

Twickenham, 16 January 1960

Wales were bursting with experience and included seven of the 1959 Lions side that had excited New Zealand critics. England, on the other hand, fielded seven new caps under a new leader, Dickie Jeeps, in a side that carried an experimental stamp. Yet within 40 minutes England had sealed the game with a performance that brimmed with confidence and adventure. Don Rutherford landed an early penalty before Mike Weston pierced the Welsh defence to send Jim Roberts speeding over for a try. Rutherford converted and kicked a penalty goal and Roberts added his second try to give the hosts a 14–0 lead at the interval – an unassailable lead in the days when Test matches were invariably low-scoring affairs in which rarely more than five points ever separated the sides.

The guiding genius behind the win was Oxford's fly half Richard Sharp, who had entered the side as an eleventh hour replacement when Bev Risman pulled a muscle. Sharp ran rings around the Welsh back row and constantly had an ordinary Welsh back division in a quandary. England went on to win the Triple Crown, only a draw in Paris blotting their Championship copybook.

STAT ZONE

Fifteen men for four Tests

The England fifteen that beat Wales in 1960 was: Don Rutherford; John Young, Malcolm Phillips, Mike Weston, Jim Roberts; Richard Sharp, Dickie Jeeps (captain); Ron Jacobs, Stan Hodgson, Peter Wright, John Currie, David Marques, Ron Syrett, Peter Robbins and Derek Morgan. The same personnel played in the 8–5 win against Ireland, the 3–all draw with France in Paris and the 21–12 win against Scotland that sealed the Triple Crown and brought a shared Five Nations title with the French. It was the first time that England had fielded the same team unchanged in a Five Nations season.

The only time since that England used the minimum fifteen players in a Championship season was for the 1991 Grand Slam. It's true that the 1995 Grand Slam starting fifteens were unaltered, but three substitutes were called on during their final match of the campaign, against Scotland. To date, England have never fielded an unchanged run-on fifteen for a Six Nations campaign.

FAMOUS CHARACTERS

The England cap-record breaker

England's most dynamic open side flanker during the 1960s was Budge Rogers. Early on he made his name as a destructive loose forward whose fitness and stamina carried him to parts of the field which lesser flankers could not reach. In his first three seasons in the England set-up the side conceded only one try to backs, a fitting tribute to Rogers's exceptional talent as a snuffer-out of attacks. That span included 1963, the best season of the Rogers years, England picking up their only Championship title of the decade.

In the mid-1960s significant changes were made to the game's laws affecting the activities of flankers. Rogers had to reassess his approach and temporarily lost his edge. Typically, though, he remained dedicated in his preparations and for fitness and stamina had no equal on the club scene. At length he developed subtle changes in his lines of defence. He was among the pioneers at perfecting a drift approach which forced opponents sideways, so stifling attacks. In addition, he still used his fitness to ensure that he was the first forward to arrive at breakdowns, but now worked on building a reputation as a constructive agent who was an invaluable asset in setting up second phase possession. He went on to captain his country and ended his distinguished international career in 1969 by breaking Wavell Wakefield's long-standing England cap record.

51

FAMOUS GAMES

Last win in Cardiff for 28 years

WALES 6, ENGLAND 13

Cardiff Arms Park, 19 January 1963

Two young and inexperienced sides faced each other in Arctic conditions yet presented the capacity crowd and a huge television audience (there being no other major outdoor sporting event to cover owing to the severe weather) with a wonderful exhibition of attacking rugby. The weather was so cold that the paint used to the mark the pitch froze and the players were issued with thermal underwear.

England went on to take the Championship title under new skipper, Richard Sharp. Forward dominance, a lively back row spearheaded by Budge Rogers, and of course Sharp to ignite a fast and talented back division were the notable ingredients in England's success.

Wales had several chances to open the scoring in the early stages but their failure to do so enabled the young visitors to grow in confidence. The first score came shortly before half-time when Mike Weston took a long throw-in on his own 25 and sent Malcolm Phillips and Peter Jackson on a breathtaking length-of-the-pitch run for Phillips to score.

In a match destined to be England's last win in Cardiff for 28 years they never looked in danger of relinquishing their stranglehold on an unusually subdued Welsh pack in the second half.

FAMOUS LASTS

The last scoreless draw

The Ireland–England match at Dublin's Lansdowne Road on 9th February, 1963, ended in a 0–0 tie and was the last of the ten pointless draws involving England since their first match in 1871. (The first 0–0 was against Scotland in Glasgow in 1873, one of six such draws with them.) The 1963 match was England's 271st cap international; their Six Nations match with Scotland in March earlier this year was their 732nd, underlining the rarity of such scorelines.

FAMOUS FIRSTS

On tour

The first British & Irish Lions tour abroad was in 1888 when Bob Seddon of England led a squad to Australia and New Zealand, but it wasn't until the summer of 1963 that an official England international team undertook a Test visit to the southern hemisphere. England, captained by the Durham centre Mike Weston, were the first individual Home Union to visit New Zealand where they were beaten 21–11 and 9–6 in a two-Test series. The touring team were also beaten 18–9 in a midweek Test against Australia in Sydney on their way home.

STAT ZONE

Pullin's run

When John Pullin returned to the England team at the start of the 1968 Five Nations, he began a run of successive appearances that lasted 36 Tests spanning seven years in the days when opportunities to win caps were fewer than in the later professional era. England typically played just four international matches a season up to the mid-eighties with only an occasional autumn international against a tour side. The players who have made the most consecutive appearances for England are:

PLAYER	RUN OF CAPS	RELEVANT SPAN
Will Carling	44	1989–1995
Jason Leonard	40	1990–1995
John Pullin	36	1968–1975
George Ford	36	2015–2018

FAMOUS CHARACTERS

The outstanding England threequarter of the early seventies

Strongly built, fast and a graceful runner who could leave defenders standing with his electric sidestep, David Duckham was the most powerful and exciting runner in the British game in the early seventies. He had no equal for sheer zest and it was his misfortune that he appeared at a time when English rugby was in the doldrums. Even so, his incisive running was an integral part of several famous England wins. True, sometimes his play lacked vision, but the element of surprise in his attacks often embarrassed opponents. The French, in particular, were left to rue their failure to deal with him in the 1973 match at Twickenham. That day Duckham thrilled the home crowd (as he had done on foreign fields for the outstanding 1971 Lions and later in Cardiff for the Barbarians) with his tricky running and collected two tries to sink French Championship hopes. At length, England went on to claim their share in a unique quintuple tie in the Championship, each of the Five Nations winning their two home matches but failing away. Later the same year he was a part of England's famous defeat of the All Blacks in Auckland and when he retired in 1976, he was

England's most-capped threequarter. If his achievements for his country look modest in retrospect, he seemed to epitomise the spirit of running rugby which uplifted British credentials in his time.

FAMOUS FIRSTS

England's first replacement player

Until the summer of 1968, replacing players in rugby matches was not permitted under International Board regulations. The game's ruling body finally conceded that teams losing a player through injury during the course of an international match should be allowed to replace him, provided medical examination confirmed that the injured player could not continue. The first replacement called on by England during a Test was Tim Dalton of Coventry against Scotland at Twickenham in 1969 when Keith Fielding sustained an ankle injury. Dalton was on the field for 50 minutes to win his only cap for England.

STAT ZONE

The most-capped subs

International teams were originally allowed to use only two replacements from a bank of four on the bench. The rules concerning bench use, however, have evolved considerably in the past 50 years so that today teams nominate eight players for reserve duties and can use them either as replacements for injured or concussed players, or as straight tactical substitutes. The five most-used England substitute/replacement players are:

Caps off the bench	Player	Career span
46	Danny Care	2008–2018
28	Mako Vunipola	2012–2019
27	Simon Shaw	1996–2011
27	Davy Wilson	2009–2015
27	Joe Worsley	1999–2011

FAMOUS FIRSTS

First replacement referee in an England Test

A peculiar variation on replacements occurred during the England–Wales Five Nations match at Twickenham in February 1970. Monsieur Robert Calmet, the French referee, was injured in the commotion of a ruck just before half-time and had to be replaced by "Johnny" Johnson, the well-known English referee who had started the match as the home team's touch-judge. It was the first time a replacement ref had been required in an England Test. Mr Johnson acted with total impartiality in his unexpected role and oversaw a Welsh win by 17–13, having taken over the whistle with England leading 13–3.

STAT ZONE

Replacement referees

England have been involved in five Tests – all at Twickenham –where the starting referee has had to be replaced:

Fixture	Year	Ref replaced	Replacement ref
England v Wales	1970	Robert Calmet (Fra)	"Johnny" Johnson (Eng)
England v France	1999	Colin Hawke (NZ)	Jim Fleming (Sco)
England v France	2001	Tappe Henning (SA)	David McHugh (Ire)
England v SA	2001	Stuart Dickinson (Aus)	David McHugh (Ire)
England v Scotland	2011	Romain Poite (Fra)	Jérôme Garcès (Fra)

STAT ZONE CENTENARY MATCH

Losing at Twickenham

England marked the centenary of the foundation of the RFU by staging a special cap match at Twickenham against an RFU President's Overseas XV, a team of leading players chosen from outside the four Home Unions. England were beaten 28–11 in front of the Queen and 50,000 of her loyal subjects. At the time it was England's biggest defeat at Twickenham but there have been five heavier reverses at the venue since then.

MARGIN OF DEFEAT	SCORE	OPPONENTS	YEAR
36	6–42	South Africa	2008
26	6–32	New Zealand	2008
21	20–41	New Zealand	2006
20	13–33	Australia	2015
18	11–29	South Africa	1997

The two heaviest losing margins were conceded on successive autumn Saturdays in 2008. England's biggest Twickenham defeat in the Five/Six Nations is 13 points: Ireland won 18–5 there in 1964.

61

FAMOUS FIRST AND LAST

The changing value of the try Part One

Growing dissatisfaction with the number of penalty goals that were deciding post-war international matches led the International Board to finally grasp the nettle and tinker with scoring values. In 1971, at the start of the northern hemisphere season, the try was upgraded from three to four points making it, for the first time in the game's history, the most valuable of all rugby's scoring actions.

Bob Hiller scored the last three-pointer for England when, in April 1971 during celebrations marking the centenary of the RFU, he crossed at Twickenham in the cap match against the RFU President's XV, a team of leading players chosen from outside the four Home Unions. Hiller in fact scored three of England's last four three-point tries, going over in the Five Nations matches earlier that year against France and Scotland. At the time he was the first full back to score tries for England for 91 years.

The scorer of England's first four-point try was the Richmond lock Chris Ralston, against Ireland at Twickenham in February 1972.

62

FAMOUS FIRST AND LAST

The Whitewash Years

Only twice in Five Nations history (1910 to 1999) were England completely whitewashed in the tournament. Their 1972 season opened with a 12–3 defeat by Wales at Twickenham before a month later 14 of that side went down 16–12 against Ireland. The selectors axed captain and full back Bob Hiller and made further changes for the away games with France and Scotland. Peter Dixon took over as captain as England were walloped 37–12 by France in Paris and wound up the season losing 23–9 against Scotland at Murrayfield.

Four years later, under Tony Neary's captaincy, England's year started full of hope with a convincing 23–6 victory over Australia at Twickenham on the first Saturday of January, only to be followed by their second and last whitewash, losing 21–9 against Wales (Twickenham), 22–12 against Scotland (Murrayfield), 12–13 to Ireland (Twickenham) and 9–30 against France (Paris).

England have never suffered a Six Nations whitewash.

63

FAMOUS GAMES

Beating the All Blacks in their own back yard

NEW ZEALAND 10, ENGLAND 16

Eden Park, Auckland, 15 September 1973

England had surprised the rugby world in the summer of 1972 when, only three months after their first whitewash in Five Nations history, they beat the Springboks in Johannesburg. A year later John Pullin again led England to a famous overseas victory, this time in the land of the All Black. England's tour had been hastily arranged after the cancellation of a proposed visit to Argentina. Pullin's side lost its first three provincial matches before pulling out all the stops to overwhelm New Zealand in the Test. "England played with admirable efficiency," wrote John Brooks of the *Christchurch Press*. "New Zealand showed a great lack of tactical appreciation and serious flaws in technique," he added.

England wisely played to their strengths. Controlled ten-man rugby was the order of the day from John Pullin and it paid handsome dividends in the form of three tries in the second half, after New Zealand had held a 10–6 lead at the break. Scrum-half Jan Webster was the mainspring of the English attack, posing his famous opponent, Sid Going, countless problems with darting runs that twice led to tries. Webster was the essential link with the forwards for the round of passing that led to England's first try and he created the move that led to Tony Neary's winning try five minutes from time.

STAT ZONE

The first of the few

Few England skippers emulated the feat of John Pullin who, by leading England to a 20–3 win against the Wallabies at Twickenham in November 1973, completed the unusual treble of skippering his country to wins over South Africa (1972), New Zealand (earlier in 1973) and Australia. The select England leaders who claim the distinction are:

ENGLAND CAPTAIN	**First win v SA**	**First win v NZ**	**First win v Australia**
John Pullin	1972	1973	1973
Will Carling	1992	1993	1988
Martin Johnson	2000	2002	2000

65

STAT ZONE

The early bath

The Australia–England Test in Brisbane at the end of May 1975 began with an unprecedented outburst of thuggery. Feet and fists were brutally used at the first ruck and, as a result of a further fracas at the game's first lineout, the England prop Mike Burton, who had been on the receiving end of unwarranted punishment in the earlier exchanges, was reprimanded and warned for butting an opponent. Then, moments later, when he tackled an Australian wing off the ball, Burton was ordered from the field to become the first player to be given his marching orders while playing for England in a Test. To date, five players have taken the early bath as a result of a sending off while playing for England in a Test:

Player sent off	Opponents	Venue	Year
Mike Burton	Australia	Brisbane	1975
Danny Grewcock	New Zealand	Dunedin	1998
Simon Shaw	New Zealand	Auckland	2004
Lewis Moody	Samoa	Twickenham	2005
Elliot Daly	Argentina	Twickenham	2016

FAMOUS CHARACTERS

The British bulldog

The player who more than any other was responsible for England's rugby revival in the late seventies was the British Bulldog from Fylde and Lancashire, Bill Beaumont. He came into the side in the mid-1970s and soon commanded a regular place through his technical expertise and special skills as a front-of-the line jumper.

It was the quality of his captaincy, however, that was to drive up England's rugby stock. He stepped in as leader in 1978 and quickly showed his knack for player management. Never one to bawl out his players, Beaumont set about establishing a good team spirit in the belief that mutual respect on the field grew from trust and good relations formed off it. Players warmed to his example and results steadily improved. With the nucleus of his successful Lancashire county side he led the Northern Division to a famous victory over the 1979 All Blacks before his rebuilding work with England reached its peak in 1980. A national side bursting with the confidence instilled by Beaumont carried off its first Grand Slam for 23 years.

He remained as captain for another two years until advised on medical grounds to take early retirement. It was a disappointing way for one of England's finest rugby servants to depart from the game he deeply loved. A firm believer in putting something back into the game, he served English and world rugby in a variety of ways afterwards and was given a knighthood in 2019, only the fourth Englishman to be honoured exclusively for services to the game.

67

FAMOUS CHARACTERS

Always in the right place at the right time

Time and again the player who turned matches in England's favour during the successful early 1990s was the Leicester flyer, Rory Underwood. A surge of pace and the instinctive genius for popping up in the right place at the right time would enable him to take a vital pass and dash over in the corner for a match-breaking try. Arguably the most important he scored was the one against France that clinched the 1991 Grand Slam. The forwards had made inroads into French territory before Simon Hodgkinson shipped a pass to Underwood wide out on the left. With plenty still to do, the wing accelerated effortlessly on a wide arc around his opponent to score in the corner.

Underwood announced his retirement in 1992, a decision he revoked a couple of months later in order to return to England colours to line up with his brother Tony on England's other wing. It was a happy partnership that lasted three years through to the 1995 Grand Slam, when England played their best rugby of the decade. When he finally retired in 1996, Rory Underwood had collected numerous England Test records with the unassuming modesty that rugby followers expect of their heroes. Even so, his 49 tries in a period that included three Grand Slams was an England career record that was richly deserved.

STAT ZONE

The top try-scorers

Rory Underwood looks likely to retain the English try-scoring record for some years to come. The leading try-getters for England in cap matches are:

Career Tries	Player	Tests	Span
49	Rory Underwood	85	1984–1996
31	Will Greenwood	55	1997–2004
31	Ben Cohen	57	2000–2006
30	Jeremy Guscott	65	1989–1999
28	Jason Robinson	51	2001–2007

Jonny May and Chris Ashton, the leading England try-getters among the current crop of players, need to more than double their count to overtake Underwood.

69

FAMOUS CHARACTERS

The reliable Mr Andrew

Every successful side needs a reliable performer. That was the role Rob Andrew filled in England's honour-laden seasons of the early 1990s when they won three Grand Slams and reached the World Cup final in 1991. He stamped his influence on the England team early in his career, dropping a goal in the first minutes of his debut international in 1985 and finishing that match with 18 points. By 1990 he had evolved into a tactical controller who, behind a winning pack, showed vision and sound judgement. There were critics of his style. Many felt he was too defensive and that his great rival for the fly-half spot, Stuart Barnes, could better kick-start the threequarter line. Occasionally Barnes took his chances to shine in the white shirt, but it was always to Andrew that the selectors turned when difficult decisions were to be made.

Towards the end of his Test days Andrew became the first in a line of deadly accurate England place kickers. His 396 points stood as the England Test record until surpassed by Jonny Wilkinson, the man Andrew groomed for stardom in his professional role as Director of Rugby at Newcastle Falcons.

FAMOUS FIRST

England's first penalty try

It wasn't until 1888 that the rugby laws made allowance for a try to be awarded by way of a penalty. The conversion, however, had to be taken on a line through where the referee adjudged that the try would have been scored. The award of a penalty try between the posts, no matter where the offence occurred did not come into force until 1937, and it was 80 years later that the kick at goal after a penalty try was finally abolished.

The first penalty try awarded to England in a Test was by Welsh referee Clive Norling in their match with Ireland in 1986, the day of Dean Richards' Test debut. The Leicester No 8 scored two tries and would have had a debut hat-trick but for a stray Irish boot kicking the ball from his toes and out of the scrum just as he swooped to pick up and dive over.

71

FAMOUS CHARACTERS

The Catalyst

The hooker whose mental steel and all-round skills epitomised the new breed of England forward in the late 1980s was Brian Moore. For years English forwards had been viewed as gentle giants by their opponents. Moore and the packs he fronted in the Carling era were tough-minded with a hatred of losing. From an early stage in his career a strike against his head or a take against his line-out throw, one sensed, were painful losses for Moore. Once he became a fixture in the Test side, he went on to play a significant role in England's resurgence. Rory Underwood, Will Carling and Rob Andrew were the public figureheads of the Grand Slam years of the early 1990s, but the key to England's successes was forward dominance, and that's where Moore figured prominently. He was the catalyst who urged the pack into a unit capable of subduing any opposition scrum. For sheer energy and determination he had no peer, whether leading by example or exhorting his colleagues to raise their game. He managed to hold his eight together at critical moments, invariably snapping their efforts into Championship-winning performances. His other great assets were courage and stamina. As England's first-choice hooker for eight years he was never dropped and, remarkably, never once required replacement in a Test match.

72

STAT ZONE

Rugby World Cup record

England are the only northern hemisphere nation that has won the Rugby World Cup. They have also reached the final on two other occasions and were beaten semi-finalists once. In 2015, they failed to progress beyond the pool stages of the tournament for the first time.

1987 – quarter-final exit, losing 16–3 to Wales in Brisbane.

1991 – finalists, losing 12–6 against Australia at Twickenham.

1995 – semi-final exit, losing 45–29 to New Zealand at Newlands, Cape Town. Lost the third/fourth place play-off 19–9 against France in Pretoria.

1999 – quarter-final exit, losing 44–21 to South Africa in Paris (after losing 30–16 against New Zealand at Twickenham in the pool stages).

2003 – winners, beating Australia 20–17 after extra time in Sydney.

2007 – finalists, losing 15–6 to South Africa in Paris.

2011 – quarter-final exit, losing 19–12 to France in Auckland.

2015 – pool stage exit, losing 28–25 against Wales and 33–13 to Australia, both at Twickenham.

73

FAMOUS CHARACTERS

The face of English rugby in the early nineties

The public face of the English rugby successes between 1988 and 1996 undoubtedly belonged to its long-serving captain, Will Carling. It was his polished threequarter play that propelled him into an England side that was given little chance of beating France in Paris in 1988. Yet new manager Geoff Cooke's young bloods went down by only a point and Carling held his place in a Five Nations season of mixed fortunes, sharpening his defensive qualities and acclimatising to the extra pace of Test rugby. By the end of the year, having established himself as a world-class player with a flair for the outside break, he was installed as captain, becoming the youngest man for 57 years to hold the honour for England. He led the side to a 28–19 victory against Australia and never looked back after that.

A psychology student, Carling did not follow the tub-thumping school of leadership. Instead he worked closely with first Geoff Cooke and later Jack Rowell to engender in his sides a collective responsibility to adhere to game plans worked out at squad training. A purposeful, friendly atmosphere developed around a side that went on to claim four Triple Crowns, three Grand Slams and reached the World Cup final in 1991 and the semi-final four years later. Carling's final record read 44 wins from 59 matches as captain. There can be no better testimony to the effectiveness of his approach.

74

STAT ZONE

Leading skippers

When Will Carling was handed the England captaincy by team manager Geoff Cooke in November 1988 he was the youngest man to lead the red rose since Peter Howard in 1931. Carling, who remains England's youngest post-war skipper, also heads the list of their captains:

Tests	Captain	Captaincy Span
59	WDC Carling	1988 to 1996
43	CDC Robshaw	2012 to 2017
39	MO Johnson	1998 to 2003
30	DM Hartley	2012 to 2018
22	LBN Dallaglio	1997 to 2004
21	WB Beaumont	1978 to 1982
21	SW Borthwick	2008 to 2010
17	ME Corry	2005 to 2007
15	PJ Vickery	2002 to 2008
13	WW Wakefield	1924 to 1926
13	NM Hall	1949 to 1955
13	E Evans	1956 to 1958
13	REG Jeeps	1960 to 1962
13	JV Pullin	1972 to 1975

The figures for Robshaw and Hartley include Tests for which they were officially named co-captain

75

FAMOUS CHARACTERS

The prince of centres

The player whose inventiveness and silky skills placed him planes above his worldwide rivals in the outside centre position for the decade from 1989 was Jeremy Guscott. He hit the England side running with four tries on his debut as a stand-in for the injured Will Carling against Romania in Bucharest in 1989 and went on to form with Carling an outstandingly succcssful midfield partnership in three Grand Slam wins. Although his defensive qualities were sometimes questioned early in his career, Guscott had no superior as an attacker. He could run and kick exquisitely, but it was for his creation of openings for others, most notably Rory Underwood, that he won the hearts of Twickenham crowds. He could rip through packed defences before invariably completing a move by giving a perfectly timed scoring pass. Some tend to forget, too, that his clear-cut breaks brought him a staggering 30 tries in an England shirt. A self-confessed rebel as a teenager, Guscott grew into an assured and easy-going character outside rugby. He said he owed much to Bath, his only club during a career that often saw him targeted by rugby league scouts. Certainly the role models at the club helped mould a personality that had an iron will to succeed both on and off the pitch.

76

FAMOUS CHARACTERS

The English Centurion

After England hooker Brian Moore had stood shoulder to shoulder with his new loose-head prop Jason Leonard for 80 minutes in Buenos Aires in 1990, he said: "It was patently obvious that he's going to be an outstanding player." Leonard was, at 21, the youngest prop to play for England since Nick Drake-Lee (in 1963). Yet even in those relatively carefree days at the start of the 1990s, when players enjoyed a few beers after a match and went back on Monday mornings to their jobs outside rugby, it was clear that England had found a genuine grafter who would take some moving in the front row. He went on to become the first and only Englishman to date to win 100 caps for his country and was the world's then most-capped forward.

His dedication to training and fitness and discipline in adhering to the strict diet regimen that are all part and parcel of the professional era enabled Leonard to stay at the top of his game during a decade that saw the demands on leading players change beyond all recognition. Throughout it all, moreover, he was a constant in England's equation. Four times he featured in Grand Slam sides (1991, 1992, 1995 and 2003), making him the first player for 75 years (and only the fourth of all time in the history of the Championship) to achieve such a distinction.

STAT ZONE

The leading cap winners

There is clear water between Jason Leonard and the next most-capped England players. The top ten are:

Caps	**Name**	**Career span**
114	Jason Leonard	1990–2004
97	Dylan Hartley	2008–2018
91	Jonny Wilkinson	1998–2011
85	Rory Underwood	1984–1996
85	Lawrence Dallaglio	1995–2007
85	Dan Cole	2010–2019
85	Ben Youngs	2010–2019
84	Martin Johnson	1993–2003
84	Danny Care	2008–2018
78	Joe Worsley	1999–2011

78

FAMOUS GAMES

Winner takes all

ENGLAND 21, FRANCE 19

Twickenham, 16 March 1991

This was only the fourth winner-takes-all Grand Slam showdown in Championship history. England, managed by Geoff Cooke and captained by Will Carling, had been denied the Grand Slam by Scotland the previous season. But on this occasion the side showed its depth of character by responding positively to a crushing blow early in the match. Simon Hodgkinson, who had nudged England ahead with an early penalty, missed with another attempt ten minutes later. But, as the English prepared for the customary drop-out to restart the game, they were caught napping by a bold and extraordinary counter-attack launched from the French dead-ball-line. It was a sublime move that covered the length of the pitch before culminating in a try for Philippe Saint-André.

Slowly but surely, however, the hosts responded. The forwards were immense in the scrums, line-outs and loose, and the composure of the entire team in the face of breathtaking French attacks was the vital quality underpinning England's confidence. England recovered to lead 18–9 by the interval, a score that included a Rory Underwood special when he arced past Jean-Baptiste Lafond to finish a slick move from second-phase possession set up by the tireless Mike Teague. In the second half, France launched a succession of frantic attacks in a bid to steal victory. Through sheer guts and a little luck, however, England hung on to achieve the first title of the Carling era.

FAMOUS FIRST AND LAST

The changing value of the try Part Two

The International Board inflated the value of a try to five points in the middle of 1992. The Preston Grasshoppers lock Wade Dooley, winning his 50th cap, scored the last four-pointer for England when, in March 1992, he crossed for the third and last try of their 24–0 win against Wales at Twickenham. The scorer of England's first five-point try was new cap Ian Hunter, breaking tackles on a 22-metre run to the line against Canada in October the same year. England won 26–13 in a game staged at Wembley Stadium while Twickenham was being refurbished.

STAT ZONE

Back-to-back Grand Slams

England are the only team that has won back-to-back Grand Slams in the International Championship three times. They first achieved the feat in the two seasons before the First World War:

1913

Beat Wales	12–0	Cardiff
Beat France	20–0	Twickenham
Beat Ireland	15–4	Dublin
Beat Scotland	3–0	Twickenham

1914

Beat Wales	10–9	Twickenham
Beat Ireland	17–12	Twickenham
Beat Scotland	16–15	Edinburgh
Beat France	39–13	Paris

WJA "Dave" Davies, succeeded by Wavell Wakefield, led the nation to another double win in the twenties:

1923

Beat Wales	7–3	Twickenham
Beat Ireland	23–5	Leicester
Beat Scotland	8–6	Edinburgh
Beat France	12–3	Paris

1924

Beat Wales	17–9	Swansea
Beat Ireland	14–3	Belfast
Beat France	19–7	Twickenham
Beat Scotland	19–0	Twickenham

And Will Carling, educated at Sedbergh (like Wakefield, his famous predecessor as captain), was in charge the last time England won back-to-back honours:

1991

Beat Wales	25–6	Cardiff
Beat Scotland	21–12	Twickenham
Beat Ireland	16–7	Dublin
Beat France	21–19	Twickenham

1992

Beat Scotland	25–7	Edinburgh
Beat Ireland	38–9	Twickenham
Beat France	31–13	Paris
Beat Wales	24–0	Twickenham

Wales (once, in 1908 and 1909) and France (once, in 1997 and 1998) are the only other teams who have done the double. No side has achieved the feat since the Six Nations era began in 2000.

FAMOUS CHARACTERS

The Great Enforcer

For strength of character and sheer physical power, it is doubtful whether English rugby will see the like of Martin Johnson again. Johnson made his Test debut against France in 1993 and quickly became a fixture in the England packs under first Geoff Cooke and then Jack Rowell, forming with Martin Bayfield the engine-room of the pack that laid the foundations of the 1995 Grand Slam. Strong and imperturbable, Johnson was the backbone of the scrum and an effective enforcer who revelled in the frolics of the loose. But he will be remembered above all for his unbending leadership. In 1997 he led the Lions to a series victory in South Africa, but it wasn't until Lawrence Dallaglio was stripped of the captaincy in the spring of 1999 that Johnson was handed the skipper's armband for England. It was the most inspired decision of Clive Woodward's reign as manager. Johnson led his country to eight consecutive wins in 2000–01, but his absence from the Grand Slam showdown in Dublin in 2001 was the telling factor in England's abject defeat. Still, his finest hours were to come back in Dublin and in Sydney in 2003 when, with Johnson at the helm, England finally achieved their first Six Nations Grand Slam before winning international rugby's ultimate prize, the Rugby World Cup.

FAMOUS FIRST

Through the card

The first English player to go through the card of scoring actions in a single match was Rob Andrew. The stand-off scored a try, landed two conversions and five penalty goals, and slotted a dropped goal to fire England to a 32–15 win against South Africa at Loftus Versfeld, Pretoria, in 1994. All told, the scoring feat has been achieved six times for England:

Scorer	Opponents	Venue	Year
Rob Andrew	South Africa	Pretoria	1994
Paul Grayson	Scotland	Murrayfield	1998
Jonny Wilkinson	Wales	Twickenham	2002
Jonny Wilkinson	New Zealand	Twickenham	2002
Charlie Hodgson	South Africa	Twickenham	2004
Jonny Wilkinson	Scotland	Twickenham	2007

FAMOUS CHARACTERS

The ultimate professional

The England player who above all typified the professionalism and determination of the 2003 Rugby World Cup-winning side was Jonny Wilkinson. The finest all-round fly half in world rugby at the time, he was already known as a deadly accurate place-kicker whose name dominated the International Championship's book of records – most points by a player in a Championship season (89 in 2001), holder of the individual match record with 35 points against Italy the same year, and the leading points scorer in the history of English Test rugby, overtaking his club mentor at Newcastle, Rob Andrew. But it was his readiness to put his body on the line that overshadowed all his brilliant kicking and tactical direction, especially in that wonderful year for England in 2003. Fly halves down the years have not exactly been renowned for their bravery in defence. Yet in match after match for England he put in the big-hitting tackles that demoralise opponents and inspire forwards.

Wilkinson won the first of his 28 caps at the age of 18 years, 314 days as a replacement wing for Mike Catt against Ireland in 1998, making him the youngest England cap since

1927. He spent a season playing at centre before becoming England's first-choice at fly half in 1999. Injuries ravaged his career for five seasons after the 2003 Rugby World Cup, but he returned to help England to another World Cup Final in 2007 and when he retired from the England scene after the 2011 event, he had amassed a record 1,179 points for his country, a haul that included most career conversions, penalties and dropped goals for the men in white.

STAT ZONE

The 300 club

Five players have scored more than 300 points in England Tests. The leading points-scorers are:

Career Points	Player	Tests	Span
1,179	Jonny Wilkinson	91	1998–2011
785	Owen Farrell	70	2012–2019
400	Paul Grayson	32	1995–2004
396	Rob Andrew	71	1985–1997
301	Toby Flood	60	2006–2013

STAT ZONE

Record defeats

At the end of his first season as England head coach, Clive Woodward took an inexperienced team missing several of his leading players on a Test tour of Australia, New Zealand and South Africa. The itinerary included four Tests in four weeks against the Tri-Nation powers, England suffering four heavy defeats on a visit described as "The Tour of Hell". Their 0–76 reverse in the opening international, against Australia in Brisbane on 6th June 1998, is the biggest score conceded and the heaviest losing margin suffered to date. England's worst defeats are:

Margin	Score	Opponents	Venue	Year
76	0–76	Australia	Brisbane	1998
48	10–58	South Africa	Bloemfontein	2007
42	22–64	New Zealand	Dunedin	1998
36	0–36	South Africa	Paris	2007
36	6–42	South Africa	Twickenham	2008

STAT ZONE

The unique dropped goal

All told, England have scored nearly 150 dropped goals in their 700-plus Tests since 1871. Only one, however, has been kicked by a forward: that was by the Leicester flanker Neil Back against Italy in Rome's Stadio Flaminio in 2000.

STAT ZONE

Record margins

England ran riot against Romania at Twickenham in the 2001 autumn internationals, scoring 20 tries, converting 14 and kicking two penalty goals in a 134–0 victory. It stands as their biggest win. The five times they have run up a century of points also comprise their five biggest winning margins:

Margin	**Score**	**Opponents**	**Venue**	**Year**
134	134–0	Romania	Twickenham	2001
110	110–0	Netherlands	Huddersfield	1998
98	106–8	United States	Twickenham	1999
98	111–13	Uruguay	Brisbane	2003
91	101–10	Tonga	Twickenham	1999

STAT ZONE

Romania 2001

That record win against Romania at Twickenham in 2001 was also the day that Charlie Hodgson, making his cap debut while Jonny Wilkinson was rested, set the record for most points by an English player in an international match. The Sale fly half scored two tries and slotted 16 goals – two penalties and 14 conversions – in a 44-point haul. Seven England players have scored 30 or more points in a single Test:

Points	Player	Opponents	Venue	Year
44	Charlie Hodgson	Romania	Twickenham	2001
36	Paul Grayson	Tonga	Twickenham	1999
35	Jonny Wilkinson	Italy	Twickenham	2001
32	Jonny Wilkinson	Italy	Twickenham	1999
30	Rob Andrew	Canada	Twickenham	1994
30	Paul Grayson	Netherlands	Huddersfield	1998
30	Jonny Wilkinson	Wales	Twickenham	2002

89

FAMOUS GAMES

Winning the Rugby World Cup

ENGLAND 20, AUSTRALIA 17

Telstra Stadium, Sydney, 22 November 2003

This was English rugby's unforgettable day, and arguably English sport's finest hour since the day in July 1966 that their football team lifted the FIFA soccer World Cup. The match was a replay of the World Cup Final a dozen years earlier when Australia had triumphed 12–6 at Twickenham. This time, victory again went to the "away" side in a tightly-contested match that was Sir Clive Woodward's 75th in charge.

Martin Johnson led the side in his last international match, Jason Robinson crossed for their only try with a first-half score in the corner and fly half Jonny Wilkinson kicked three penalties. For Australia, Lote Tuqiri scored their only try when gathering a high kick by Stephen Larkham to open the scoring in the sixth minute of the first half. Elton Flatley matched Wilkinson's penalty goals and a match played in a downpour was tied at 14–all after normal time.

There followed the ultimate test of nerve: 20 minutes of extra time on a rain-swept night. But cometh the hour, cometh the men. Johnson's leadership and belief in his team's ability to succeed was never more obvious than in the prolongations. Wilkinson and Flatley exchanged penalties before in the closing moments the English fly half, as tough mentally and every bit as composed as his skipper, picked his moment to slot the winning dropped goal and send a nation into ecstasy.

STAT ZONE

The dropped goal king

It's no exaggeration to say that there will never be a more important dropped goal for England than Jonny Wilkinson's to win the 2003 Rugby World Cup. It was one of the record 36 he kicked for his country during his career, a world record. The most dropped goals kicked for England are:

Drops	Player	Caps	Career
36	Jonny Wilkinson	91	1998–2011
21	Rob Andrew	71	1985–1997
6	Paul Grayson	32	1995–2004
4	John Horton	13	1978–1984
4	Les Cusworth	12	1979–1988
4	Andy Goode	17	2005–2009

91

STAT ZONE

Putting in a full shift

The only player who featured throughout England's 580 field minutes of pool stages, quarter-, semi- and final in the successful 2003 Rugby World Cup campaign was the Wasps back-row, Lawrence Dallaglio. He put in an unprecedented full shift that included the 20 minutes of overtime needed to clinch the Webb Ellis Trophy in extra time in Sydney.

It was the constructive and forthright play that he brought to England's 1993 Rugby World Cup Sevens triumph that catapulted Lawrence Dallaglio into the rugby limelight. Having almost achieved a full set of England representative honours, he subsequently broke into the England squad after the 1995 Rugby World Cup. He proved such an outstanding loose forward that he quickly became the player England could not afford to drop. After Phil de Glanville relinquished his brief hold on the captaincy in succession to Will Carling, Dallaglio was given the skipper's armband when Clive Woodward entered the scene in November 1997. Dallaglio went on to lead his country 14 times before Martin Johnson took over the captaincy against Australia in June 1999. During Dallaglio's time at the helm, England enjoyed a Triple Crown season in

ENGLAND
RUGBY
WORLD CUP
2003

1998 and later the same year brought to an end South Africa's world record-equalling run of 17 successive Test victories.

A long-term shoulder injury and a knee injury that brought a premature end to his Lions tour in Australia in 2001 disrupted his progress. But in 2003, after starting the Championship campaign on the bench against France, he found new reserves of strength and determination to reclaim his place alongside Neil Back and Richard Hill in a back-row that played together more times at Test level than any other international trio. Dallaglio was at his best in Dublin in 2003, where he scored the opening try of the winner-takes-all Grand Slam showdown with Ireland, and later that year continued that form into the World Cup campaign. One of English rugby's most respected servants, he finally stood down from England duty after playing in the losing 2007 Rugby World Cup Final against South Africa in Paris.

STAT ZONE

Longest losing streak

It's hard to believe that England's record losing streak was equalled when they were reigning world champions. Sir Clive Woodward had departed as head honcho two years earlier when, in 2006, England embarked on a miserable run of seven successive losses that heralded the end of Andy Robinson's stewardship as manager:

Date	Result	Opposition	Venue
25th Feb 2006	Lost 12–18	Scotland	Murrayfield
12th Mar 2006	Lost 6–31	France	Stade de France
18th Mar 2006	Lost 24–28	Ireland	Twickenham
11th Jun 2006	Lost 3–34	Australia	Sydney
17th Jun 2006	Lost 18–43	Australia	Melbourne
5th Nov 2006	Lost 20–41	New Zealand	Twickenham
11th Nov 2006	Lost 18–25	Argentina	Twickenham

Twice before, between 1904 and 1906, and 1971 and 1972, the national side had experienced the woe of seven defeats on the trot.

STAT ZONE

Mind the gap

Kevin Yates of Bath won his second England cap against Argentina in Buenos Aires in June 1997, but had to wait until May 2007 before winning his third cap, against South Africa at the Free States Stadium in Bloemfontein. He was a member of the Saracens club by then but the nine years and 353 days gap between successive caps set a record for an England player.

O2

94

STAT ZONE

England's longest-serving international player

Simon Shaw holds the record for the longest span between making his England debut and bowing out from the national side. He won his first cap against Italy in a 54–21 win at Twickenham in November 1996, and finished his England service 14 years and 319 days later in the 2011 Rugby World Cup quarter-final defeat against France.

RUGBY
2007

FAMOUS GAMES

Beating the world champions

ENGLAND 38, NEW ZEALAND 21

Twickenham, 1 December 2012

England's fans could hardly believe it. New Zealand, the reigning world champions, went into the match unbeaten in 20 Tests. But in the most significant performance of Stuart Lancaster's tenure and Chris Robshaw's captaincy, England set new records for their highest score and biggest winning margin against New Zealand. Thanks to sturdy defence and the trusty boot of Owen Farrell, England steadily built their lead in the first half and had three penalties and a dropped goal to show for their efforts with nothing conceded. Another Farrell penalty extended the lead to 15 points early in the second half before a typical All Blacks riposte almost wiped out the margin. Tries in quick succession by Julian Savea and Kieran Read, both converted by Dan Carter, meant that suddenly only one point separated the sides. Would England buckle? The capacity crowd need not have worried for Robshaw inspired his team to respond quickly and efficiently, Brad Barritt, Chris Ashton and Manu Tuilagi running in tries in a purple patch that extended England's lead to 32–14. Two late Freddie Burns penalties and a consolation try for Savea that was converted by Aaron Cruden completed the scoring.

STAT ZONE

England's oldest try-scorer

When England beat Uruguay 60–3 at the Manchester City Stadium in October 2015 in their last match of the Rugby World Cup, they crossed for ten tries, including a hat-trick by veteran No 8 Nick Easter. He was 37 years and 56 days old when he scored, making him the oldest English try-scorer to date. He actually broke a record he already held, having scored (aged 36½) coming off the bench against Italy at Twickenham in the Six Nations earlier the same year. The only other England player to score a Test try after his 36th birthday was Eric Evans, against France in 1957.

STAT ZONE

Longest winning streak

The immediate reaction to England's disappointing exit at the pool stages of the 2015 Rugby World Cup was to replace Stuart Lancaster with the former Australia and Japan coach, Eddie Jones. Jones embarked on a wonderful run of 17 victories as England's head honcho, an unprecedented start for any Tier One coach. Taken with the last win of Lancaster's reign, it represents England's longest run of successive Test victories and equalled the Tier One world record (18) set by New Zealand between 2015 and 2016. England's winning run:

Won 60–3	v Uruguay 2015 (h)
Won 15–9	v Scotland 2016 (a)
Won 40–9	v Italy 2016 (a)
Won 21–10	v Ireland 2016 (h)
Won 25–21	v Wales 2016 (h)
Won 31–21	v France 2016 (a)
Won 27–13	v Wales 2016 (h)
Won 39–28	v Australia 2016 (a)
Won 23–7	v Australia 2016 (a)

Won 44–40	v Australia 2016 (a)
Won 37–21	v South Africa 2016 (h)
Won 58–15	v Fiji 2016 (h)
Won 27–14	v Argentina 2016 (h)
Won 37–21	v Australia 2016 (h)
Won 19–16	v France 2017 (h)
Won 21–16	v Wales 2017 (a)
Won 36–15	v Italy 2017 (h)
Won 61–21	v Scotland 2017 (h)

STAT ZONE

The record score draw

It's generally regarded as the most unsatisfactory result in rugby, but from time to time international matches end drawn, and occasionally they turn out to be memorable affairs as when England and Scotland shared 76 points in a wonderfully topsy-turvy match at Twickenham in March 2019. It was the 51st drawn international that England had been involved in – more than any other rugby-playing nation – and the highest score draw involving any Tier One rugby nation. England's first tie was a scoreless affair against Scotland in Glasgow in 1873 in their third international match, and the 38–all result in 2019 was their 19th against the Scots. England's highest scoring draws are:

Score	Opponents	Year	Venue
38–38	Scotland	2019	Twickenham
26–26	New Zealand	1997	Twickenham
19–19	Argentina	1981	Buenos Aires
15–15	Australia	1997	Twickenham
15–15	Scotland	2010	Murrayfield
14–14	Wales	1904	Leicester
14–14	France	1971	Twickenham
14–14	South Africa	2012	Port Elizabeth
13–13	Wales	1983	Cardiff
12–12	France	1974	Paris
12–12	Scotland	1989	Twickenham

THE HEAD HONCHO

England's head coaches

It was in the autumn of 1969 that England adopted a squad system for the first time. Thirty players were named in August to take part in a series of monthly practices under the direction of the Northampton and England flanker, Don White, as preparation for the December Test against South

Africa at Twickenham. The experiment proved a resounding success, a team containing five new caps under the captaincy of Bob Hiller achieving its first victory over the Springboks (11–8).

Head Coach	**Tenure**	**P**	**W**	**D**	**L**	**Win rate**
Don White	1969–71	11	3	1	7	27%
John Elders	1972–74	16	6	1	9	38
John Burgess	1975	6	1	0	5	17
Peter Colston	1976–79	18	6	1	11	33
Mike Davis	1979–82	16	10	2	4	63
Dick Greenwood	1983–85	17	4	2	11	24
Martin Green	1985–87	14	5	0	9	36
Geoff Cooke	1987–94	49	35	1	13	71
Jack Rowell	1994–97	29	21	0	8	72
Clive Woodward	1997–2004	83	59	2	22	71
Andy Robinson	2004–06	22	9	0	13	41
Brian Ashton	2007–08	22	12	0	10	55
Rob Andrew	2008	2	0	0	2	0
Martin Johnson	2008–11	38	21	1	16	55
Stuart Lancaster	2012–15	46	28	1	17	61
Eddie Jones	2016–19	40	31	1	8	78

RUGBY IN THE BLOOD

Since the first Test in 1871 (when the Stokes, Luscombe and Birkett families were represented) to 2019 (when the Farrell, Youngs and Vunipola families featured), players have followed fathers and brothers into the England national side.

Fathers and sons

Reg (first capped in 1871) and John (first capped in 1906) **Birkett**
Andy (2007) and Owen (2012) **Farrell**
John (1981) and Rob (1998) **Fidler**
Dick (1966) and Will (1997) **Greenwood**
Frank (1899) and Reg (1932) **Hobbs**
George (1892) and John (1930) **Hubbard**
Rob (1984) and Alex (2017) **Lozowski**
William (first capped in 1874) and Jumbo (1904) and Cecil (1906) **Milton**
Ivor (1948) and Peter (1972) **Preece**
Frank (1907) and Keith (1947) **Scott**
Bill sen (1894) and Bill jun (1926) **Tucker**
Harry (1901) and Bill (1933) **Weston**

Harry sen (1889) and Harry jun (1929) **Wilkinson**
Nick (1983) and Ben (2010) and Tom (2012) **Youngs**
Fe'ao (first capped by Tonga in 1988) was the father of Mako (first capped by England in 2012) and Billy (first capped by England in 2013) **Vunipola**
George (first capped by England in 1905) was the father of Walter (first capped by Wales in 1938) **Vickery**

Brothers

Delon (first capped in 2008) and Steffon (first capped in 2009) **Armitage**
Reginald (1871) and Louis (1875) **Birkett**
Henry (1874) and Charles (1876) **Bryden**
Frederick (1894) and Frank (1897) **Byrne**
James (1897) and Joseph (1899) **Davidson**
Wyndham (1875) and Arthur (1883) **Evanson**
George (1947) and Nigel (1954) **Gibbs**
Harry (1875) and John (1876) **Graham**
George (1899), Charles (1901) and Thomas (1905) **Gibson**
Temple (1878) and Charles (1880) **Gurdon**
William (1876), Robert (1880) and James (1882) **Hunt**
Richard (1925) and Tom (1928) **Lawson**
John (1871) and Francis (1872) **Luscombe**
Stephen (1929) and George (1934) **Meikle**
John (1904) and Cecil (1906) **Milton**
Charles (1910) and Robert (1914) **Pillman**
Mason (1887) and William (1889) **Scott**
Aubrey (1882) and Norman (1886) **Spurling**
Frederic (1871) and Lennard (1875) **Stokes**

Adrian (1905) and Frederick (1910) **Stoop**
Frank (1897) and Percy (1898) **Stout**
John (1929) and Deneys (1931) **Swayne**
Frederick (1914) and Frank (1920) **Taylor**
Henry (1879) and Arthur (1883) **Taylor**
Edward (1876) and George (1876) **Turner**
Rory (1984) and Tony (1992) **Underwood**
Mako (2012) and Billy (2013) **Vunipola**
Harold (1936) and Arthur (1937) **Wheatley**
Ben (2010) and Tom (2012) **Youngs**

George (first capped for Ireland in 1878) was the brother of Arthur (first capped for England in 1887) **Fagan**
Frank (first capped for Wales in 1885) was the brother of Froude (first capped for England in 1886) **Hancock**
Manu **Tuilagi** (first capped for England in 2011) has five brothers capped by Samoa

STAT ZONE

The international record

England (with Scotland) are the oldest nations on the international circuit. Up to 31st March 2019, they had played 732 international matches, won 403, drawn 51 and lost 278.

Opponents	First cap match	P	W	D	L
Argentina	1981	23	18	1	4
Australia	1909	50	24	1	25
Canada	1992	6	6	0	0
Fiji	1988	7	7	0	0
France	1906	105	58	7	40
Georgia	2003	2	2	0	0
Ireland	1875	134	77	8	49
Italy	1991	25	25	0	0
Japan	1987	2	2	0	0
Netherlands	1998	1	1	0	0
New Zealand	1905	41	7	1	33
New Zealand Natives	1889	1	1	0	0
Pacific Islands	2008	1	1	0	0
RFU President's XV	1971	1	0	0	1

Romania	1985	5	5	0	0
Samoa	1995	8	8	0	0
Scotland	1871	137	75	19	43
South Africa	1906	42	15	2	25
Tonga	1999	2	2	0	0
United States	1987	5	5	0	0
Uruguay	2003	2	2	0	0
Wales	1881	132	62	12	58
TOTAL	**1871**	**732**	**403**	**51**	**278**

The record against Australia includes the 1928 match against New South Wales subsequently given full Test status by the Australian Rugby Union.

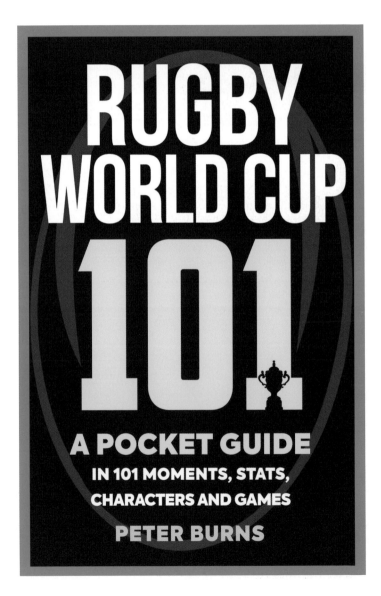

ALSO AVAILABLE FROM

POLARIS

PUBLISHING

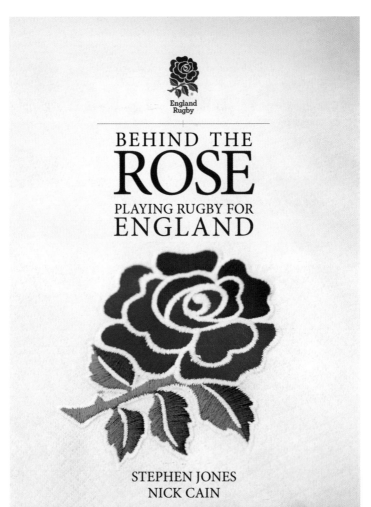

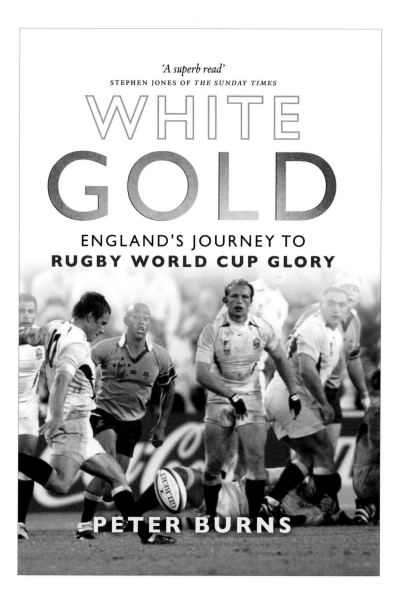

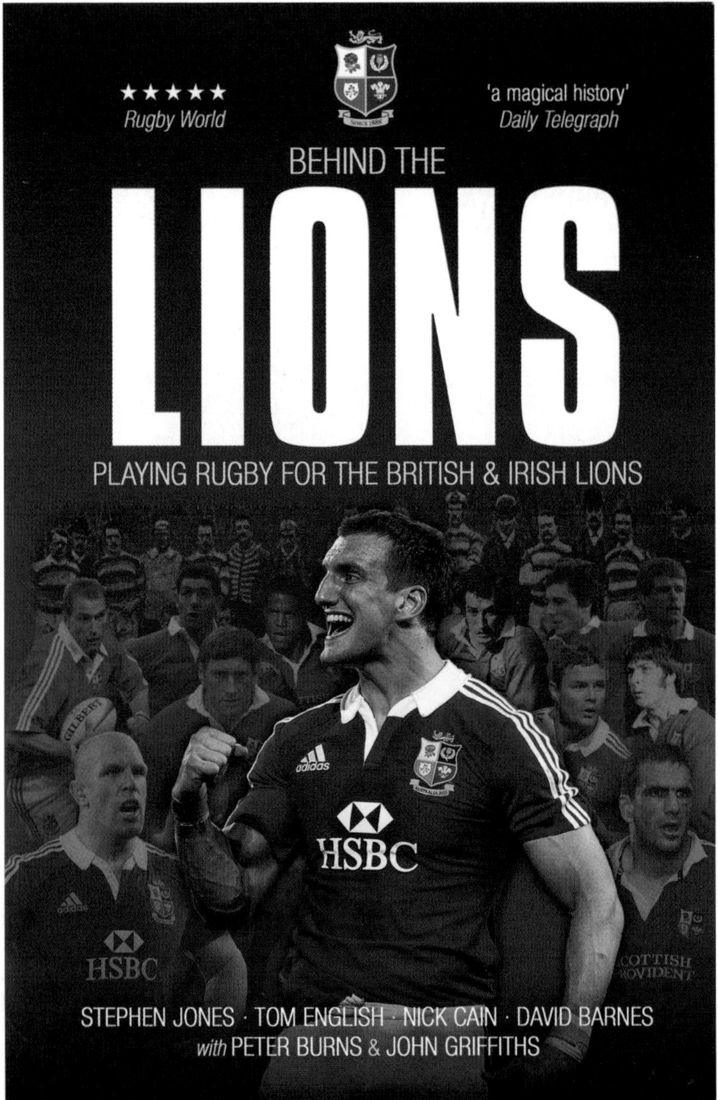
★★★★★
Rugby World
'a magical history'
Daily Telegraph
BEHIND THE
LIONS
PLAYING RUGBY FOR THE BRITISH & IRISH LIONS
STEPHEN JONES · TOM ENGLISH · NICK CAIN · DAVID BARNES
with PETER BURNS & JOHN GRIFFITHS

ALSO AVAILABLE FROM

POLARIS

PUBLISHING

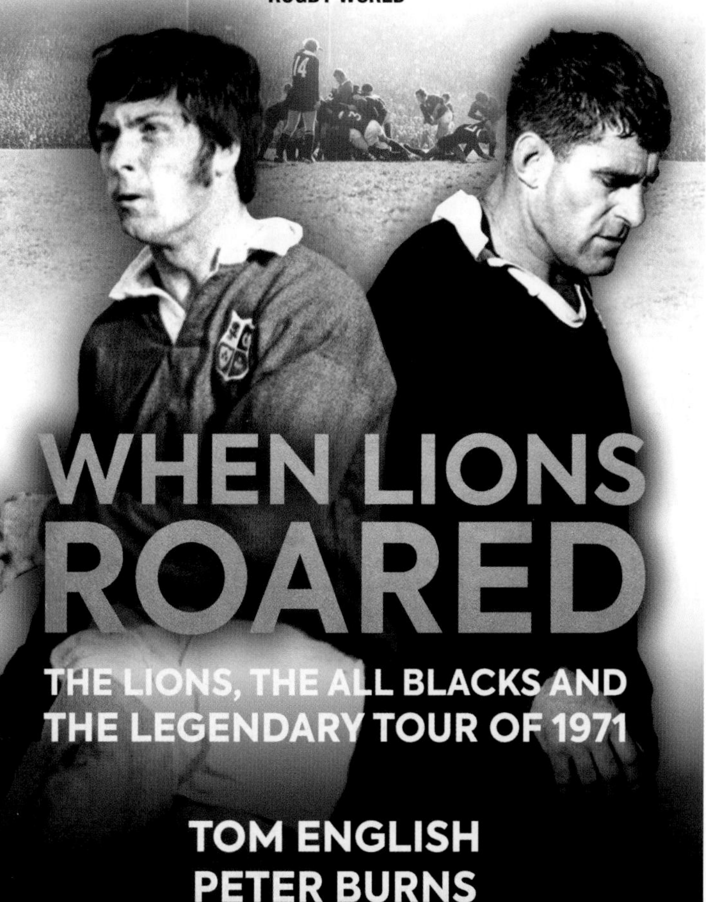

POLARIS
PUBLISHING